Productivity and Efficiency Measurement of Airlines

Productivity and Efficiency Measurement of Airlines

Data Envelopment Analysis using R

Boon L. Lee
Associate Professor, School of Economics and Finance,
Faculty of Business and Law, QUT, Australia

ELSEVIER

Elsevier
Radarweg 29, PO Box 211, 1000 AE Amsterdam, Netherlands
The Boulevard, Langford Lane, Kidlington, Oxford OX5 1GB, United Kingdom
50 Hampshire Street, 5th Floor, Cambridge, MA 02139, United States

Notices

Knowledge and best practice in this field are constantly changing. As new research and experience broaden our understanding, changes in research methods, professional practices, or medical treatment may become necessary.

Practitioners and researchers must always rely on their own experience and knowledge in evaluating and using any information, methods, compounds, or experiments described herein. In using such information or methods they should be mindful of their own safety and the safety of others, including parties for whom they have a professional responsibility.

To the fullest extent of the law, neither the Publisher nor the authors, contributors, or editors, assume any liability for any injury and/or damage to persons or property as a matter of products liability, negligence or otherwise, or from any use or operation of any methods, products, instructions, or ideas contained in the material herein.

ISBN: 978-0-12-812696-7

For information on all Elsevier publications visit our website at
https://www.elsevier.com/books-and-journals

Publisher: Joseph P. Hayton
Acquisitions Editor: Kathryn Eryilmaz
Editorial Project Manager: Aleksandra Packowska
Production Project Manager: Kiruthika Govindaraju
                            & Selvaraj Raviraj
Cover Designer: Vicky Pearson Esser

Typeset by TNQ Technologies

Working together
to grow libraries in
developing countries

www.elsevier.com • www.bookaid.org

To

Sue Chuein and Larc

Contents

Preface

This book is designed to provide researchers and academics a range of R programming scripts and the use of R packages available from CRAN (the Comprehensive R Archive Network) to measure efficiency and productivity via data envelopment analysis (DEA). This book was inspired from the authors' experience and anecdotal comments on why there does not exist a book that provides explicit instructions via step-by-step use on R programming scripts to measure efficiency and productivity. While this book provides a wide range of R programming scripts and the use of specific R packages on DEA, it is by no means complete as it is impossible to consider all DEA models.

This book may be used as a textbook or as a reference text to accompany other DEA books. This book is extremely helpful as it provides the application side to the understanding of DEA. Undergraduates and postgraduates can use the R programming scripts or R packages detailed in the book and work through the models and generate efficiency and productivity estimates using the sample data provided in the book.

This book is not designed to provide a thorough theoretical background of the various DEA models and not designed to compete with the existing excellent DEA books. There are already plenty of such books available. For example Färe et al. (1985), Färe et al. (1994), Coelli et al. (1998), Cooper et al. (2000), Coelli et al. (2005), Cooper et al. (2007) and O'Donnell (2018). In fact, this book aims to complement all other DEA studies, theoretical and applied, by providing a step-by-step guide on the use of specific DEA models with the associated R programming scripts.

Another purpose of this book is to contextualize the efficiency and productivity performance of international airlines through the use of DEA models. The focus on measuring the efficiency and productivity of airlines is because of deregulation of the global airline industry with the aim of maximizing efficiency. By employing DEA (i.e. standard DEA and its variants), this book aims to identify and demonstrate the adoption of the appropriate DEA model in order to measure the efficiency and productivity of airlines and explain their performance since deregulation.

We believe this book offers readers something unique like a one-stop shop (or book) because the book provides a range of explicit R programming scripts and a step-by-step use on the measurement of efficiency and productivity. We hope that this book also provides a point of departure for others to further develop and improve the existing DEA models.

Chapter 1

Introduction

1.1 Introduction

The purpose of this book is to provide data envelopment analysis (DEA) practitioners DEA models written in the R programming script to measure the efficiency and productivity of airlines (or any other industry for that matter). This book is not meant to provide a thorough theoretical background of the various data envelopment analysis (DEA) models as there are already plenty of such books available, for example, Färe et al. (1985), Färe et al. (1994), Coelli et al. (1998), Cooper et al. (2000), Coelli et al. (2005), Cooper et al. (2007) and O'Donnell (2018). This book aims to complement all other DEA studies, theoretical and applied, by providing a step-by-step guide on the use of specific DEA models with the associated R programming scripts, and interpretation of the results therefrom. By making it user-friendly, we hope that DEA practitioners are able to apply the models with minimal fuss. As the number of DEA models is extensive, this book does not cover all of them. Only the key DEA models are covered to meet the needs of most DEA practitioners.

Another purpose of this book is to measure the productivity and efficiency performance, albeit using a sample of international airlines. The focus on measuring the efficiency and productivity of airlines is because of deregulation of the global airline industry with the aim of maximizing efficiency. By employing DEA (i.e., standard DEA and its variants), this book aims to identify and demonstrate the adoption of the appropriate DEA model to measure the efficiency and productivity of airlines and explain their performance since deregulation.

This chapter is divided into four sections. Following the introduction in Section 1.1, we provide a brief comment on the evolution of the global airline industry since deregulation in Section 1.2. Section 1.3 presents a brief history of the developments of the DEA model and evolution of variants of DEA. This chapter concludes in Section 1.4 with an outline of the chapters of this book.

1.2 Evolution and deregulation of the global airline industry—a brief comment

The Airline Deregulation Act of 1978 (ADA) and its international counterpart, International Air Transportation Act of 1979 (IATA), were major turning

Productivity and Efficiency Measurement of Airlines. https://doi.org/10.1016/B978-0-12-812696-7.00008-8

points in the US airline industry. Prior to the ADA, the airline industry was heavily regulated, and leading economists, such as Alfred E. Kahn, had argued that regulation creates inefficiencies and leads to higher costs. The ADA opened up domestic competition in the United States with airlines adopting competitive and operating strategies such as price competition (i.e., fare reductions) and adopting hub-and-spoke route systems. The ADA also opened up competition in the form of entrance of low-cost carriers (LCCs) into the market. The 'Open Skies' agreement was also introduced during this period, and it was only in 1992 when the first European country, the Netherlands, signed the first Open Skies agreement with the United States. Morrison and Winston (1986) noted that the financial outcomes from the deregulation of the US airline industry were also an impetus to the European Union's air transport liberalization. As observed in Button (2001, p.255), 'by 1987 passenger enplanements had risen by 88% compared with 1976, employment in the industry had risen from 340,000 to 450,000, scheduled passenger miles were up by 62% and seat availability had grown by 65%'.

The deregulation process of the European Union (EU) air transport began in 1987 and comprised three packages (Button, 2001). The 'First Package' started in January 1988 introduced a degree of flexibility in the aviation sector. The 'Second Package' in November 1990 included further flexibility on fares, capacity restrictions and market access. The 'Third package' in January 1993 removed the remaining barriers of aviation, thus culminating in the formal EU airline liberalization.

Besides the EU, many nations also began deregulating their airline industry since the deregulation of the US airline industry. Canada began deregulating in 1979 and was fully deregulated from 1988 (Oum et al., 1991). Deregulation in Japan began in 1986 with the enactment of the revised Aviation Law in February 2000 culminating in full liberalization (Ida and Tamura, 2005). Australia began deregulating the airline industry in the late 1970s and was fully deregulated in 1990 after the repeal of the 'Two Airline agreement' under the Airline Agreement Termination Act, which saw the abandonment of fare setting and limits on entry to the industry (Quiggin, 1997). China began airfare deregulation in 1997 through the adoption of price discrimination (Wang et al., 2016). Airline deregulation began in Korea with the Airline Deregulation Act in 2008 (Kim, 2016).

The process of air transport deregulation in ASEAN (Association of South-East Asian Nations) progressed with each milestone of signing of agreement/ declaration towards full liberalization. This began with the Bangkok Summit Declaration of 1995, which included the development of an Open Sky Policy (Forsyth et al., 2006). In 2001, the 7th Air Transport Ministers' Meeting resulted in the regional initiative for the progressive and phased liberalization of air services in ASEAN. In 2004, the 10th Air Transport Ministers' Meeting formed the Action Plan for ASEAN Air Transport Integration and Liberalization 2005–2015 (Laplace and Latgé-Roucolle, 2016). In 2009, the

ASEAN Multilateral Agreement on Air Services was signed in Manila, Philippines. This was followed up with the 2010 Multilateral Agreement for the Full Liberalization of Passengers Air Services signed in Brunei Darussalam. These agreements culminated with the operation of the ASEAN Single Aviation Market in 2016, also known as the ASEAN Open Skies Agreement (Heriyanto and Putro, 2016).

The airline industry also witnessed the development of alliances among airlines, mainly to overcome various regulatory and financial obstacles (Kottas and Madas, 2018). Since 1997, the three key global airline alliances, Oneworld, SkyTeam and Star alliance, have dominated the global airline industry. The growth of these alliances is shown in Fig. 1.1.

As a result of the events that have occurred in the global airline industry, namely deregulation leading to intense competition with the entrance of LCCs; and the creation of the global alliances, the aforementioned events provide a unique opportunity to apply the nonparametric model, DEA and its variants to measure the efficiency and productivity of airlines in the era of deregulation.

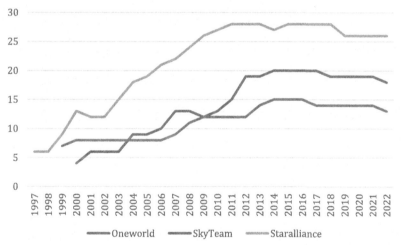

FIGURE 1.1 Number of airlines per global airline alliance, 1997—2022. *Source: Trend diagram based on data drawn from Wikipedia. https://en.wikipedia.org/wiki/Oneworld, https://en. wikipedia.org/wiki/SkyTeam https://en.wikipedia.org/wiki/Star_Alliance.*

1.3 A brief history of developments in data envelopment analysis

The theory behind deregulation is to achieve maximum efficiency (Keeler, 1984; Peltzman et al., 1989). To that end, this book uses the nonparametric model, DEA, to measure the efficiency and productivity performance of airlines. DEA, based on linear mathematical programming, constructs a piecewise frontier from a given data set. It essentially measures the efficiency of decision-making units (DMUs) relative to the frontier. The frontier efficiency concept was first elucidated in Farrell's (1957) seminal work, but the term DEA was first introduced in Charnes et al. (1978), henceforth CCR. The CCR model, under constant returns to scale assumption, was further developed by Banker et al. (1984), by introducing a variable returns-to-scale DEA model. Since then, DEA has been employed in numerous studies covering a range of industries. For a bibliography of DEA studies, we refer readers to Seiford (1997), Tavares (2002), Gattoufi et al. (2004a, 2004b), Cook and Seiford (2009), Emrouznejad and Yang (2018) and Contreras (2020).

While DEA gained popularity because of its simplicity and applicability, it nonetheless has several limitations (Stolp, 1990; Coelli et al., 1998). Most of these limitations can be overcome by following the protocols described in Dyson et al. (2001) and Cook et al. (2014). There are, however, two limitations that are much more serious. As noted in Schmidt (1986), there is no error term in DEA indicating any error may well be attributed to inefficiency. Also DEA efficiency scores have no statistical significance due to its nonparametric nature (Grosskopf, 1996). The lack of statistical inference has since drawn considerable interests among practitioners of DEA to overcome these challenges by incorporating sensitivity analysis using statistical tests (Sengupta, 1987; Valdmanis, 1992) and bootstrapping (Simar and Wilson, 1998).

DEA has evolved to accommodate and reflect a range of production models that the standard DEA is unable to. For example, variants of DEA models were developed to reflect the production models that incorporate both good (desirable) and bad (undesirable) outputs. Halkos and Petrou (2019) provide a critical review of a list of DEA studies that incorporate bad output and Section 2.5, which focuses only on airline studies. Then, there are other DEA variants that reflect a network process whereby outputs produced in the first stage of production become inputs in the following stage of production. Such models were developed because the standard DEA treats the production process as a 'black box' and simply transforms inputs into outputs and neglects any possible intervening processes, including dissimilar series or parallel functions. This problem is overcome by using a network DEA (NDEA) model developed by Färe (1991) and Färe and Grosskopf (1996, 2000). For a list of bibliography on NDEA, we refer readers to Koronakos (2019) and Section 2.6, which focuses only on airline studies.

DEA has also evolved as an applied tool for consulting purposes or for policy analysis (Grosskopf, 1996). This involved a two-stage procedure whereby DEA efficiency scores estimated in the first stage are regressed against a host of independent variables (sometimes referred to as explanatory variables or environmental variables) in the second stage intended to explain inefficiency. The most common regression models used in DEA studies are ordinary least squares (OLS), Tobit and Simar and Wilson (2007) bootstrap truncated regression. Lovell et al. (1995) argued that OLS was inappropriate due to censoring issues, and this led to Tobit regression being widely used in DEA two-step analysis. Simar and Wilson (2007) found that many two-stage studies suffer from the problem of serial correlation and the lack of a coherent data-generating process. Hence, they extended this area of research by proposing a bootstrap truncated regression model based on the maximum likelihood estimation.

1.4 Outline of chapters

This book comprises nine chapters. Following the introduction in Chapter 1, a literature review of airline studies using DEA and variants of DEA is covered in Chapter 2. Chapter 3 introduces the standard DEA model, the cost and revenue efficiency models and the accompanying R scripts to these models. Testing of the reliability of the results and the associated R scripts are also covered. Chapter 4 focuses on the productivity growth models that measure productivity change over time. These include the Malmquist productivity index, Hicks−Moorsteen productivity index, Lowe productivity index and Färe−Primont productivity index. The accompanying R scripts to these models are also included in the chapter. Chapter 5 describes other variants of DEA measurement, such as the metafrontier DEA, slacks-based measure, superefficiency DEA, potential gains DEA and directional distance function, and their R scripts. Chapter 6 covers some variants of DEA that incorporate bad outputs in the model and their R scripts. Chapter 7 looks at some network DEA models and includes their scripts. Chapter 8 describes the second-stage regression models, namely OLS, Tobit and Simar and Wilson (2007) bootstrap truncated regression model, and includes their R scripts. This book concludes in Chapter 9 with some brief remarks.

References

Banker, R.D., Charnes, A., Cooper, W.W., 1984. Some models for estimating technical and scale inefficiencies in data envelopment analysis. Management Science 30, 1078−1092.

Button, K., 2001. Deregulation and liberalization of European air transport markets. Innovation: The European Journal of Social Science Research 14 (3), 255−275. https://doi.org/10.1080/13511610120102619.

Charnes, A., Cooper, W.W., Rhodes, E., 1978. Measuring the efficiency of decision making units. European Journal of Operational Research 2 (6), 429−444.

Coelli, T., Rao, D.S.P., Battese, G.E., 1998. An Introduction to Efficiency and Productivity Analysis. Kluwer Academic Publishers, Boston.

Coelli, T., Rao, D.S.P., O'Donnell, C.J., Battese, G.E., 2005. An Introduction to Efficiency and Productivity Analysis, second ed. Springer, Boston.

Contreras, I., 2020. A review of the literature on DEA models under common set of weights. Journal of Modelling in Management 15, 1277−1300.

Cook, W.D., Seiford, L.M., 2009. Data envelopment analysis (DEA) - thirty years on. European Journal of Operational Research 192, 1−17.

Cook, W.D., Tone, K., Zhu, J., 2014. Data envelopment analysis: prior to choosing a model. OMEGA 44, 1—4.

Cooper, W.W., Seiford, L.M., Tone, K., 2000. Data Envelopment Analysis. Kluwer Academic Publishers.

Cooper, W.W., Seiford, L.M., Tone, K., 2007. Data Envelopment Analysis A Comprehensive Text with Models, Applications, References and DEA-Solver Software, second ed. Springer, New York.

Dyson, R.G., Allen, R., Camanho, A.S., Podinovski, V.V., Sarrico, C.S., Shale, E.A., 2001. Pitfalls and protocols in DEA. European Journal of Operational Research 132, 245—259.

Emrouznejad, A., Yang, G.-L., 2018. A survey and analysis of the first 40 years of scholarly literature in DEA: 1978—2016. Socio-Economic Planning Sciences 61, 4—8.

Färe, R., 1991. Measuring Farrell efficiency for a firm with intermediate inputs. Academia Economic Papers 19, 329—340.

Färe, R., Grosskopf, S., 1996. Intertemporal Production Frontiers: With Dynamic DEA. Kluwer Academic Publishers, Boston.

Färe, R., Grosskopf, S., 2000. Network DEA. Socio-Economic Planning Science 34, 35—49.

Färe, R., Grosskopf, S., Lovell, C.A.K., 1985. The Measurement of Efficiency of Production. Kluwer Academic Press, Boston.

Färe, R., Grosskopf, S., Lovell, C.A.K., 1994. Production Frontiers. Cambridge University Press, Cambridge.

Farrell, M.J., 1957. The measurement of productive efficiency. Journal of the Royal Statistical Society: Series A 120 (3), 253—290.

Forsyth, P., King, J., Rodolfo, C.L., 2006. Open skies in ASEAN. Journal of Air Transport Management 12, 143—152.

Gattoufi, S., Oral, M., Reisman, A., 2004a. A taxonomy for data envelopment analysis. Socio-Economic Planning Sciences 38, 141—158.

Gattoufi, S., Oral, M., Reisman, A., 2004b. Data envelopment analysis literature: a bibliography update (1996—2001). Socio-Economic Planning Sciences 38, 122—159.

Grosskopf, S., 1996. Statistical inference and nonparametric efficiency: a selective survey. Journal of Productivity Analysis 7, 161—176.

Halkos, G., Petrou, K.N., 2019. Treating undesirable outputs in DEA: a critical review. Economic Analysis and Policy 62, 97—104.

Heriyanto, D., Putro, Y., 2016. Challenges and opportunities in the establishment of ASEAN open skies policy. PADJADJARAN Jurnal Ilmu Hukum (Journal of Law) 6, 466—488.

Ida, N., Tamura, T., 2005. Effects of deregulation on local air passenger demand. Research in Transportation Economics 13, 211—231.

Keeler, T.E., 1984. Theories of regulation and the deregulation movement. Public Choice 44, 103—145.

Kim, D., 2016. The effects of airline deregulation: a comparative analysis. Asian Journal of Business Environment 6, 5—10. https://doi.org/10.13106/eajbm.2016.vol6.no3.5.

Koronakos, G., 2019. A taxonomy and review of the network data envelopment analysis literature. In: Tsihrintzis, G., Virvou, M., Sakkopoulos, E., Jain, L.C. (Eds.), Machine Learning Paradigms. Learning and Analytics in Intelligent Systems, vol. 1. Springer, Cham. https://doi.org/10.1007/978-3-030-15628-2_9.

Kottas, A.T., Madas, M.A., 2018. Comparative efficiency analysis of major international airlines using data envelopment analysis: exploring effects of alliance membership and other operational efficiency determinants. Journal of Air Transport Management 70, 1—17.

Laplace, I., Latgé-Roucolle, C., 2016. Deregulation of the ASEAN air transport market: measure of impacts of airport activities on local economies. Transportation Research Procedia 14, 3721–3730.

Lovell, C.A.K., Walters, L.C., Wood, L.L., 1995. Stratified models of Education production using Modified DEA and regression analysis. In: Charnes, A., Cooper, W.W., Lewin, A.Y., Seiford, L.M. (Eds.), Data Envelopment Analysis: Theory, Methodology and Applications. Kluwer, Boston, pp. 329–352.

Morrison, S., Winston, C., 1986. The Economic Effects of Airline Deregulation. Brookings Institution, Washington.

Oum, T.H., Stanbury, W.T., Tretheway, M.W., 1991. Airline deregulation in Canada and its economic effects. Transportation Journal 30, 4–22.

O'Donnell, C.J., 2018. Productivity and Efficiency Analysis: An Economic Approach to Measuring and Explaining Managerial Performance. Springer, Singapore.

Peltzman, S., Levine, M.E., Noll, R.G., 1989. The Economic Theory of Regulation after a Decade of Deregulation. Brookings Papers on Economic Activity (Microeconomics), pp. 1–59.

Quiggin, J., 1997. Evaluating airline deregulation in Australia. The Australian Economic Review 30, 45–56.

Schmidt, P., 1986. Frontier production functions. Econometric Reviews 4, 289–328.

Seiford, L.M., 1997. A bibliography for data envelopment analysis (1978–1996). Annals of Operations Research 73, 393–438.

Sengupta, J., 1987. Data envelopment analysis for efficiency measurement in the stochastic case. Computers & Operations Research 14, 117–129.

Simar, L., Wilson, P.W., 1998. Statistical analysis of efficiency scores: how to bootstrap in nonparametric frontier models. Management Science 44, 49–61.

Simar, L., Wilson, P.W., 2007. Estimation and inference in two-stage, semi parametric models of production processes. Journal of Econometrics 136, 31–64.

Stolp, C., 1990. Strengths and weaknesses of data envelopment analysis: an urban and regional perspective. Computers, Environment and Urban Systems 14, 103–116.

Tavares, G., 2002. A Bibliography of Data Envelopment Analysis (1978–2001). RUTCOR, Rutgers University. RRR 01–02. Available at: https://mathsci2.appstate.edu/wmcb/Class/5340/ClassNotes141/DEA_Bib.pdf.

Valdmanis, V., 1992. Sensitivity analysis for DEA models: an Empirical example using Public vs. NFP Hospitals. Journal of Public Economics 48, 185–205.

Wang, J., Bonilla, D., Banister, D., 2016. Air deregulation in China and its impact on airline competition 1994–2012. Journal of Transport Geography 50, 12–23.

Chapter 2

Literature on data envelopment analysis in airline efficiency and productivity

2.1 Introduction

This chapter provides a description of the current literature on the efficiency and productivity performance of airlines in terms of the model, the sample size and years covered, the number of airlines, and the inputs and outputs (and environmental variables where applicable). The literature on airline efficiency and productivity performance can be divided into six groups. The first group uses the standard DEA model to measure efficiency performance of airlines. The second group measures cost and profit efficiency of airlines. The third group of studies measures the productivity performance of airlines over a period of time using productivity and efficiency change models such as Malmquist productivity and Luenberger productivity. The fourth group of studies incorporates bad outputs into the model via either DEA or slacks-based measure DEA model to measure the environmental efficiency of airlines. The fifth group of studies uses the network DEA or dynamic DEA model. The sixth group looks at studies that employ other DEA variations. The seventh group looks at studies that includes a second-stage regression.

The purpose of this chapter is not to provide a comprehensive discussion of results from each study. The purpose is mainly to identify the differences in the choice of models and variables, and acknowledge the rationale for the model assumptions. It is important to note that not all studies provide their rationale and as such, we only mention those that include their rationale.

The chapter is divided into nine sections. Following the introduction in Section 2.1, the literature on airline efficiency and productivity studies based on the standard DEA model are covered in Section 2.2. Section 2.3 covers airline performance studies based on cost and profit efficiency. Section 2.4 covers the airline studies that employ productivity change models. Section 2.5 covers the airline studies that incorporate bad outputs in their studies to measure environmental efficiency. Section 2.6 covers airline performance studies that employ network DEA or dynamic DEA. Section 2.7 covers airline performance studies

Productivity and Efficiency Measurement of Airlines. https://doi.org/10.1016/B978-0-12-812696-7.00006-4

that employ other variations of the DEA model. Section 2.8 covers airline studies that include a second-stage regression analysis. This chapter concludes in Section 2.9 with brief comments.

2.2 Literature on airline efficiency using standard data envelopment analysis model

The earliest airline efficiency studies employ the radial DEA model. Radial here refers to the proportionate movements, which are covered in more detail in Chapter 3. These studies generally assume either constant returns to scale (CRS) or variable returns to scale (VRS). DEA under CRS was first introduced by Charnes et al. (1978) also known as CCR and then followed by Banker et al. (1984) with VRS model known as BCC. Table 2.1 shows the list of airline efficiency studies that employed the standard DEA model.

The vast majority of airline studies use a number of employees to reflect labour and considered as an input, which is easily discernible. What is not discernible is the airline capacity, namely available ton (tonne) kilometres/miles or available seat kilometres/miles. Chan and Sueyoshi (1991), Schefczyk (1993), Sengupta (1998), Fethi et al. (2000), Scheraga (2004, 2006), Merkert and Hensher (2011), Merkert and Morrell (2012), and Jain and Natarajan (2015) used this variable as an input, whereas Distexhe and Perelman (1994), Greer (2006), Bhadra (2009), Assaf and Josiassen (2011), Wang et al. (2011), Lee and Worthington (2014), Nolan et al. (2014), and Wang et al. (2019) used this as an output. The underlying difference between the two groups of studies is the production process. If airline capacity is considered an input, the outputs usually relate to some revenue indicator, for example, revenue passenger kilometers/miles, nonpassenger revenue, and profitability. If airline capacity is considered an output, then the inputs are usually the number of aircraft, aircraft fuel, total assets, and operating expenses. Because each study focuses an aspect of the airline production process, the standard DEA is somewhat limited in its ability to truly measure the airline performance. This limitation has since been resolved with the development of the network DEA model, and discussion of this is detailed in Section 2.6.

Schefczyk (1993) considered CRS by assuming that airlines have the opportunity to influence their own scale over a period of time. Good et al. (1995) and Tofallis (1997) considered CRS because of the vast majority of airline literature discussed in White (1979) that supported the CRS assumption via statistical analysis. Alam and Sickles (1998, 2000) used CRS based on their test which was not rejected and also found that Cornwell et al. (1990) and Caves et al. (1984) supported the use of CRS. Sickles et al. (2002) used CRS as their statistical test does not reject it and that the CRS assumption is consistent with the vast majority of the airline literature. In similar fashion, Hu et al. (2017) used CRS after their statistical test rejected the VRS while the CRS estimators were consistent. Min and Joo (2016) use CRS because it

TABLE 2.1 Airline efficiency studies based on standard DEA.

Author(s)	Model	Sample	Inputs and outputs
Chan and Sueyoshi (1991)	DEA (input orientation, CRS)	30 US airlines, 1973–84	Inputs: Available ton km, operating cost, nonflight assets. Outputs: Profitability (operating profit (loss) divided by total operating revenues), growth (annual percentage change in revenue passenger miles).
Schefczyk (1993)	DEA (input orientation, CRS)	15 international airlines, 1990	Inputs: Available ton km, operating cost, nonflight assets. Outputs: Revenue passenger km, nonpassenger revenue.
Distexhe and Perelman (1994)	DEA (input orientation, CRS)	33 international airlines, 1977–88	Inputs: Labour (flying personnel), number of aircraft, network density (average number of aircraft departures per 100,000 km). Outputs: Available ton km, weight load factor (ton km performed divided by available ton km).
Good et al. (1995)	DEA (input orientation, CRS)	16 European and American airlines, 1976–86	Inputs: Labour index (pilots/other cockpit crew, flight attendants, mechanics, ticket/passenger handlers, other personnel), materials index (fuel, nonflight equipment, other materials), flight equipment. Outputs: Aggregate revenue output (passenger services, cargo services, incidental services).
Charnes et al. (1996)	DEA (multiplicative model)	10 Latin American airlines, 1988	Inputs: Seat kilometres available, cargo-ton kilometres available, fuel, labour. Outputs: Passenger kilometres performed.
Ray and Mukherjee (1996)	DEA (input orientation, CRS)	21 US airlines, 1983–84	Inputs: Labour (index), fuel (quantity index), materials (quantity index), flight equipment (quantity index), ground equipment. Outputs: Quantity index of four outputs (revenue passenger miles [RPM] of scheduled flights, RPM of chartered flights, revenue ton miles [RTM] of mail, RTM of other freight).

Continued

TABLE 2.1 Airline efficiency studies based on standard DEA.—cont'd

Author(s)	Model	Sample	Inputs and outputs
Tofallis (1997)	DEA (input orientation, CRS and VRS)	14 international airlines, 1990	Inputs: aircraft capacity in ton km, operating cost, nonflight assets. Outputs: Passenger km, nonpassenger revenue.
Alam and Sickles (1998)	DEA (output orientation, CRS)	11 US airlines, 1970–90	Inputs: flight capital (number of planes), labour (pilots, flight attendants, mechanics, passenger and aircraft handlers, and other labour), energy (aircraft fuel), materials (supplies, outside services, and nonflight capital). Outputs: Quantity of revenue output (revenue ton miles) including passenger and nonscheduled revenue output.
Sengupta (1998)	DEA (input orientation, CRS)	14 international airlines, 1990	Inputs: available ton kilometre, operating cost, total nonflight assets. Outputs: revenue passenger kilometre, non passenger revenue.
Alam and Sickles (2000)	DEA (input orientation, CRS)	11 US airlines, 1970–90	Inputs: flight capital (number of planes), labour (pilots, flight attendants, mechanics, passenger and aircraft handlers, and other labour), energy (aircraft fuel), materials (supplies, outside services, and nonflight capital). Outputs: Quantity of revenue output (revenue ton miles) including passenger and nonscheduled revenue output.
Fethi et al. (2000)	DEA (input orientation, CRS)	17 European airlines, 1991–95	Inputs: Available ton km, operating cost, nonflight assets. Outputs: Revenue passenger km, nonpassenger revenue ton km.
Sickles et al. (2002)	DEA (input orientation, CRS)	16 European airlines, 1977–90	Inputs: Total number of employees, the number of aircraft, scheduled network size (in route kilometers), proportion of jet aircraft, proportion of wide bodied jet aircraft. Outputs: Total revenue output, capacity output.

Study	Method	Sample	Inputs/Outputs
Capobianco and Fernandez (2004)	DEA (input orientation, VRS)	170 international airlines, 1993–97	Inputs: Financial leverage (ratio of total assets to equity). Outputs: Profitability (return on assets measured as the ratio of net profit to total assets), firm size (natural logarithm of net sales), tangibility of assets (ratio of fixed to total assets).
Desli and Ray (2004)	DEA (output orientation, VRS)	21 US airlines, 1984	Inputs: Quantity indexes of labor, quantity index of fuel, quantity index of materials, quantity index of flight equipment, quantity index of ground equipment. Outputs: Quantity index (numbers revenue passenger miles flown, ton kilometers of cargo flown, ton kilometers of mail flown).
Scheraga (2004)	DEA (input and output orientation; and nonoriented, CRS)	38 international airlines, 1995 and 2000	Inputs: Available ton km, operating cost, nonflight assets. Outputs: Revenue passenger km, nonpassenger revenue ton km.
Greer (2006)	DEA (input orientation, CRS)	14 US airlines, 2004	Inputs: Labour, aircraft fuel, fleet-wide seating capacity. Outputs: Available seat miles.
Scheraga (2006)	DEA (nonorientation and input orientation, VRS)	34 international airlines, 2003 and 2004	Inputs: Available ton km, operating cost, nonflight assets. Outputs: Revenue passenger km, nonpassenger revenue.
Barbot et al. (2008)	DEA (input orientation, VRS)	49 international airlines, 2005	Inputs: Number of employees, airline fleet, fuel, other operating inputs. Outputs: Revenue passenger km, revenue tonne km, ancillary output other than passengers and cargo (other revenues).
Barros and Peypoch (2009)	DEA (output orientation, CRS and VRS)	29 international airlines, 2000–05	Inputs: Number of employees, operational cost, number of planes. Outputs: Revenue passenger km, EBIT (earnings before interest and taxes).
Bhadra (2009)	DEA (output orientation, VRS)	13 US airlines, 1985–2006	Inputs: Gallons of jet fuel, number of full time employees (FTEs), ratio of flight stage miles to trip stage miles, utilization of aircraft (in hours), number of seats per aircraft, number of aircraft. Outputs: Available seat miles.

Continued

TABLE 2.1 Airline efficiency studies based on standard DEA.—cont'd

Author(s)	Model	Sample	Inputs and outputs
Chow (2010)	DEA (input orientation, CRS)	10 Chinese airlines, 2003–04	Inputs: Full time employee number, aircraft fuel used, seat capacity. Outputs: Revenue ton km of passengers and freight.
Hong and Zhang (2010)	DEA (input orientation, CRS)	29 international airlines, 1998–2002	Inputs: Number of employees, available seat km. Outputs: Total revenue, revenue passenger kilometre, freight RTK.
Ouellette et al. (2010)	DEA (input orientation, VRS)	7 Canadian airlines, 1960–99	Inputs: Quantity of investment (ratio of expenditures on investment (includes maintenance of equipment) to price index), fleet capital (perpetual inventory method), labour, energy, materials. Outputs: Unit toll (freight and passenger tons km of regular flights) and charter flights (tons km).
Assaf and Josiassen (2011)	DEA (output orientation, VRS)	15 UK airlines, 2002–07	Inputs: Labour expenses (total expenditure for the salaries and allowances of all employees), aircraft fuel and oil expenses, aircraft value (total financial value of aircraft minus depreciation). Outputs: Available tonne kilometres, total operational revenues.
Merkert and Hensher (2011)	DEA (input orientation, CRS and VRS)	58 international airlines, 2007/08 and 2008/09	Inputs: Available tonne kilometres, number of staff (FTE). Outputs: Revenue passenger kilometres revenue tonne kilometres.
Wang et al. (2011)	DEA (input orientation, CRS)	30 international airlines, 2006	Inputs: Number of employees, fuel expense, number of planes. Outputs: Available seat miles, revenue passenger miles, nonpassenger revenue.
Merkert and Morrell (2012)	DEA (input orientation, CRS and VRS)	66 international airlines, 2007/08 and 2008/09	Inputs: Available tonne kilometres, number of staff (FTE). Outputs: Revenue passenger kilometres, revenue tonne kilometres, total revenues.

Study	Method	Sample	Inputs/Outputs
Wu et al. (2013)	DEA (input orientation, CRS and VRS)	12 international airlines, 2006–10	Inputs: Number of full time employees, operational costs, number of operational aircraft. Outputs: Revenue tonne kilometres, operating revenue.
Lee and Worthington (2014)	DEA (output orientation, VRS)	42 international airlines, 2006–10	Inputs: Average number of employees, total assets in US dollars, kilometres flown. Outputs: Available tonne kilometres.
Nolan et al. (2014)	DEA (input orientation, CRS and VRS)	18 international airlines, 1998–2010	Inputs: Value of productive assets, number of aircraft, quantity of fuel used (in US gallons), total number of employees. Outputs: Available seat miles, the number of destinations served.
Wu and Liao (2014)	DEA (input orientation, CRS and VRS)	38 international airlines, 2006–10	Inputs: Revenue passenger kilometer (RPK), number of passengers, energy (fuel) cost, capital cost, material cost, labour cost, other operating expense per employee. Outputs: Operating revenue, return on investment, return on assets, net income.
Choi et al. (2015)	Service quality-adjusted DEA (output orientation, CRS)	12 American airlines, 2008–11	Inputs: Number of employees, available seat miles. Outputs: Revenue passenger miles, operating revenue, service quality index.
Jain and Natarajan (2015)	DEA (input orientation, VRS)	12 Indian airlines, 2006	Inputs: Available ton kilometres, operating cost. Outputs: Revenue passenger kilometres, nonpassenger revenue.
Min and Joo (2016)	DEA (output orientation, CRS and VRS)	59 international airlines, 2009	Inputs: Operating expenses, underutilization (in percentage). Outputs: Operating revenue, number of passengers, revenue passenger kilometres.
Sarangal and Nagpal (2016)	DEA (input orientation, VRS)	13 Indian airlines, 2005–06 to 2011–12	Inputs: Staff strength (the number of full time employees), available seat kilometres. Outputs: Revenue passenger kilometres.

Continued

TABLE 2.1 Airline efficiency studies based on standard DEA.—cont'd

Author(s)	Model	Sample	Inputs and outputs
Choi (2017)	DEA (input orientation, VRS)	14 US airlines, 2006 to 2015	Inputs: Cost per available seat mile. Outputs: Revenue per ASM, passenger yield, load factor.
Hu et al. (2017)	DEA (input orientation, CRS and VRS)	15 ASEAN airlines, 2010–14	Inputs: Number of aircraft, operating cost, available seat kilometers. Outputs: Revenue passenger kilometers, total revenue.
Kottas and Madas (2018)	DEA (input orientation, CRS)	30 international airlines, 2012–16	Inputs: Number of employees, total operating cost, number of operated aircraft. Outputs: Total operating revenue, revenue passenger kilometers, revenue tonne kilometres.
Sakthidharan and Sivaraman (2018)	DEA (input orientation, CRS and VRS)	Indian airlines, 2013–14	Inputs: Revenue passenger kilometres, freight tonne kilometres. Outputs: Available tonne kilometres, cost per available seat kilometre (ASK), cost of fuel per ASK, nonfuel cost, maintenance cost per ASK, ownership cost (ownership per ASK), number of employees.
Mhlanga et al. (2018)	DEA (input orientation, VRS)	7 South African airlines, 2012–15	Inputs: Number of full-time employees, available seat kilometres, operating expense less employee expenditure per available seat kilometres, employee expenditure per number of full-time employees. Outputs: Revenue passenger kilometres, operating revenue per available seat kilometres.
Hermoso et al. (2019)	DEA (output orientation, CRS and VRS)	43 European airlines, 2014	Inputs: Number of employees, total assets, number of destinations. Outputs: Sales, number of passengers.

overcomes the limitations imposed by financial data when used in ranking airlines' financial performances and operating efficiencies. Kottas and Madas (2018) considered CRS because their study aimed at assessing overall efficiency. Sakthidharan and Sivaraman (2018) assumed CRS because airlines are not restricted by regulation to a particular operating region.

Capobianco and Fernandez (2004) assumed VRS as their study deals with organizations of different sizes. Assaf and Josiassen (2011) also assumed VRS as they argue that CRS is only appropriate when airlines are operating at an optimal level of scale. Also, they argue that the market conditions in the period of their study were highly competitive suggesting that airlines may not be operating optimally. Merkert and Hensher (2011), Merkert and Morrell (2012) and Sarangal and Nagpal (2016) assumed VRS because of imperfect competition, budgetary restrictions and regulatory constraints on entries and mergers that lead to firms operating inefficiently. Lee and Worthington (2014) also share the aforementioned view and further argue the use of VRS because airlines vary in sizes. Jain and Natarajan (2015) and Sakthidharan and Sivaraman (2018) assumed VRS because airlines may be operating at different scale of operations and that airlines using a production technology may not always exhibit economies of scale.

With regards to the orientation, we observe that the choice of input or output orientation revolves around airline operations. Schefczyk (1993) considered input orientation because from an airline operations perspective, the output of an airline is a target that is specified by management. To meet this target, management will have to minimize consumption of resources (i.e., inputs). Similar to Schefczyk (1993), several studies assumed that airlines have better ability to control their inputs than outputs. These include Capobianco and Fernandez (2004), Merkert and Hensher (2011), Merkert and Morrell (2012), Sarangal and Nagpal (2016), Hu et al. (2017), Kottas and Madas (2018), Sakthidharan and Sivaraman (2018) and Mhlanga et al. (2018). Fethi et al. (2000) considered input orientation because 'in a growing competitive market for air transport the outputs are less likely to be under the control of the individual airlines than their choice of inputs'. Barros et al. (2013) assumed output orientation because the objective of the airlines is to maximize outputs.

2.3 Literature on airline cost efficiency, revenue efficiency and profit efficiency

The measurement of cost, revenue and profit efficiency allows a firm to evaluate its financial performance. Adopting a low-cost strategy may be important for some firms to achieve overall cost advantage. Considering that inputs across airlines are not entirely homogeneous and vary in prices, this can impact on a firm's cost advantage. Where revenue performance is the objective, a firm can take advantage of higher output prices given the market structure it is in by providing a service at higher prices that may be perceived as higher quality (Banker and Johnston, 1994).

From the list of studies in Table 2.2, only a handful provided rationale for their choice of returns to scale. Siregar and Norsworthy (2001) assumed VRS because the airline industry was experiencing increasing returns to scale (Pitt and Norsworthy, 1999) in part from technological change associated with the characteristics of aircrafts operated in the industry. Sarangal and Nagpal (2016) assumed VRS because of imperfect competition, budgetary restrictions and regulatory constraints on entries and mergers that lead to firms operating inefficiently. In terms of orientation, Sarangal and Nagpal (2016) considered an input orientation under the assumption that airlines have better ability to control their inputs than outputs.

2.4 Literature on airline productivity change performance

There are times when airline performance is best measured over a period of time. For example, the Airline Deregulation Act (ADA) of 1978 opened up competition in the US domestic market. Opening up of international markets, such as the North Atlantic and Pacific markets, followed suit. Liberalization of the airline market also witnessed the entrants of low-cost carriers and price competition. We can thus see that the effects of deregulation are ongoing as well as the behaviour of each airline relative to their competitors. Hence, productivity and efficiency change over time would be a better measure of airline performance. Table 2.3 presents a literature of airline performance based on various productivity change models.

Alam and Sickles (2000) used CRS based on their hypothesis test that does not reject CRS and found that Cornwell et al. (1990) and Caves et al. (1984) supported the use of CRS. Lee et al. (2017) also assumed CRS because of the airline literature stated above supporting the use of CRS. They also noted that Caves et al. (1984) empirically demonstrated that US large and small carriers could compete with one another over extended periods of time, an observation that is consistent with CRS. Sickles et al. (2002) used CRS as their statistical test does not reject it and that the CRS assumption is consistent with the vast majority of the airline literature.

2.5 Literature on airline efficiency incorporating bad output

The main activity of airlines is to convey passengers and freight. This suggests that airlines produce outputs, such as revenue passenger kilometres and tonne kilometres performed, which are considered good outputs. Unfortunately, the conveying of good outputs consumes fuel and inevitably produces greenhouse gases (e.g., CO_2 emissions), which are considered bad outputs.

TABLE 2.2 Airline efficiency studies based on DEA cost efficiency, DEA revenue efficiency and DEA profit efficiency.

Author(s)	Model	Sample	Inputs and outputs
Banker and Johnston (1994)	Cost efficiency and revenue efficiency (input and output orientation, VRS)	12 US airlines, 1981–85 (quarterly data)	Inputs: Aircraft and traffic servicing labour (ground labour handling aircraft, passengers, baggage and cargo), reservations and sales labour, flight crew labour (pilots, copilots, navigators and flight engineers), flight attendant labour, fuels and oils, maintenance labour, maintenance materials and overhead, ground property and equipment, flight equipment, general overhead. Outputs: Revenue passenger miles (RPM).
Ray and Mukherjee (1996)	Cost efficiency DEA (CRS)	21 US airlines, 1983–84	Inputs: Labour (index), fuel (quantity index), materials (quantity index), flight equipment (quantity index), ground equipment. Outputs: Quantity index of four outputs (RPM of scheduled flights, RPM of chartered flights, revenue ton miles (RTM) of mail, RTM of other freight).
Siregar and Norsworthy (2001)	DEA (output orientation, VRS)	4 US airlines, 1970–92	Inputs: Annual quantity of capital stock, annual quantities of materials, labor, energy, purchased business services, other expenses. Outputs: RPM.
Coelli et al. (2002)	Profit efficiency DEA (output orientation, VRS)	28 international airlines, 1990	Inputs: Total number of personnel, fuel consumed, other (calculated by subtracting personnel and fuel expenses from total operating expenses). Price of inputs: Annual average wages, average fuel prices, international purchasing power parities. Outputs: Passenger kilometres, tonne kilometres of freight. Price of outputs: Average fares obtained by dividing revenues by the output measures.
Sarangal and Nagpal (2016)	Cost efficiency (input orientation, VRS)	13 Indian airlines, 2005–06 to 2011–12	Inputs: Operating expense less employee expenditure per ASK, employee expenditure per full-time employee. Outputs: Operating revenue per ASK.

TABLE 2.3 Airline performance studies based on productivity change models.

Author(s)	Model	Sample	Inputs and outputs
Distexhe and Perelman (1994)	Malmquist (input orientation, CRS)	33 international airlines, 1977–88	Inputs: Labour (flying personnel), number of aircraft, network density (average number of aircraft departures per 100,000 km). Outputs: Available ton km, weight load factor (ton km performed divided by available ton km),
Ray and Mukherjee (1996)	Malmquist (input orientation, CRS)	21 US airlines, 1983–84	Inputs: Labour (index), fuel (quantity index), materials (quantity index), flight equipment (quantity index), ground equipment. Outputs: Quantity index of four outputs (revenue passenger miles (RPM) of scheduled flights, RPM of chartered flights, revenue ton miles (RTM) of mail, RTM of other freight).
Alam and Sickles (2000)	Malmquist (input efficiency, CRS)	11 US airlines, 1970–90	Inputs: flight capital (number of planes), labour (pilots, flight attendants, mechanics, passenger and aircraft handlers, and other labour), energy (aircraft fuel), materials (supplies, outside services and nonflight capital). Outputs: Quantity of revenue output (revenue ton miles).
Sickles et al. (2002)	Malmquist (input orientation, CRS)	16 European airlines, 1977–90	Inputs: Total number of employees, the number of aircraft, scheduled network size (in route kilometers), proportion of jet aircraft, proportion of wide bodied jet aircraft. Outputs: Total revenue output, capacity output.
Greer (2008)	Malmquist (input orientation, CRS)	14 US airlines, 2000–04	Inputs: Labour, aircraft fuel, fleet-wide seating capacity. Outputs: Available seat miles.
Chow (2010)	Malmquist (output orientation, CRS)	10 Chinese airlines, 2003–04; 16 Chinese airlines, 2005–07	Inputs: Full-time employee number, aircraft fuel used, seat capacity. Outputs: Revenue ton km of passengers and freight.
Assaf (2011)	Bootstrapped Malmquist productivity (input orientation, CRS)	18 UK airlines, 2004–07	Inputs: Total operational cost (excluding labour cost), labour cost, aircraft fuel, oil expenses, aircraft value. Outputs: Tonne kilometres available, total operational revenues.

Study	Method	Sample	Inputs/Outputs
Pires and Fernandes (2012)	Malmquist (input orientation, VRS)	42 international airlines, 2001–02	Inputs: Financial leverage (ratio of total assets to equity). Outputs: Firm size (natural logarithm of net sales), tangibility of assets (ratio of fixed to total assets), intangible assets.
Barros and Couto (2013)	Luenberger productivity and Malmquist index (nonoriented, CRS)	23 European airlines, 2000–11	Inputs: Number of employees, operational cost, number of seat kilometres available. Outputs: Revenue per passenger kilometres, revenue cargo tonnes carried.
Lee et al. (2015)	Malmquist–Luenberger index (output orientation, CRS)	35 international airlines, 2004–11	Inputs: Number of employees, average aircraft capacity, fuel burn, hours flown. Outputs: Tonne kilometres performed, CO_2 emissions.
Scotti and Volta (2015)	Biennial Malmquist–Luenberger index (output orientation, VRS)	18 European airlines, 2000–10	Inputs: Available seat kilometres, available freight-tonne kilometres. Outputs: Revenue passenger kilometres, total freight tonne kilometres, CO_2 emissions.
Yang and Wang (2016)	Bootstrapped Malmquist productivity (input orientation, CRS)	25 international airlines, 2001–06	Inputs: Available ton kilometres, number of planes, number of employees. Outputs: Revenue passenger kilometres, revenue ton kilometres.
Lee et al. (2017)	Luenberger productivity index (output orientation, CRS)	34 international airlines, 2004–11	Inputs: Number of employees, average aircraft capacity, fuel burn, hours flown. Outputs: Revenue tonne kilometres, CO_2 emissions.
Zhang et al. (2017)	Malmquist–Luenberger index (output orientation, CRS)	18 US and Chinese airlines, 2011–14	Inputs: Number of aircraft, labour, fuel consumption. Outputs: Revenue tonne kilometres, operating revenue, CO_2 emissions.
Wang et al. (2019)	Malmquist productivity (Orientation unspecified, returns to scale unspecified)	17 Asian airlines, 2012–16 and 2017–21.	Inputs: Total assets, number of operating aeroplanes, operating expenses. Outputs: Available seat kilometres, revenue passenger kilometres.

In the past few decades, airlines have taken major strides in reducing carbon emissions either through the use of new technologies or/and through the adoption of operational methods. These attempts, largely driven by commercial interests, include collaboration between aircraft and engine manufacturers to develop and adopt innovative technologies and high-performing products to make airlines more efficient, cost-effective, and environmentally friendly (Lee et al., 2015). To measure airline environmental performance, it is thus essential to incorporate both good and bad outputs into the production model as both outputs are associated with fuel consumption. Table 2.4 shows the literature on airline performance incorporating bad outputs.

Lee et al. (2015, 2017) assumed CRS because it is consistent with the vast majority of airline literature such as White (1979), Cornwell et al. (1990), Good et al. (1995), Alam and Sickles (2000) and Sickles et al. (2002). They also noted that Caves et al. (1984) empirically demonstrated that US large and small carriers could compete with one another over extended periods of time and found that airline production technology can be characterized as having CRS. Conversely, Cui and Li (2016) considered VRS because it is more suitable for time series analysis.

2.6 Literature on airline performance based on network DEA or DEA linked by phases

The airline studies in Section 2.2 employ standard DEA to measure technical efficiency based on the production assumption of some set of inputs used to generate some set of outputs. However, we observed from the literature that some studies consider airline capacity, namely available ton (tonne) kilometres/miles or available seat kilometres/miles, as inputs, whereas others as outputs. Close observation of this reveals that the airline production process is much more complex and comprises individual production processes. The use of the standard DEA is therefore insufficient as the standard DEA treats the production process as a 'black box' in that it simply transforms inputs into outputs and neglects any possible intervening processes, including dissimilar series or parallel functions.

To overcome this issue, the network DEA (NDEA) model was developed. NDEA, developed by Färe (1991) and Färe and Grosskopf (1996, 2000), comprise networks of 'divisions' or 'nodes'. Each division/node produces output(s), which may themselves become input(s) in the production of other outputs in subsequent divisions/nodes. Each node provides an efficiency score and when linked together provides an estimate of overall efficiency for each DMU.

Since then, NDEA has been extensively used in many studies covering a wide range of industries, including the airline industry. Table 2.5 presents the literature on airline performance studies that employ NDEA or DEA linked by phases.

TABLE 2.4 Airline performance studies incorporating bad outputs.

Author(s)	Model	Sample	Inputs and outputs
Chang et al. (2014)	Two-stage SBM-NDEA (Nonoriented, CRS and VRS)	27 international airlines, 2010	Inputs: Available ton kilometres, fuel consumption, number of employees. Outputs: Revenue ton kilometres, profits, CO_2 emissions.
Cui and Li (2015)	Virtual Frontier Benevolent DEA Cross Efficiency (input orientation, CRS)	11 international airlines, 2008–12	Inputs: Number of employees, capital stock, aviation kerosene (tons). Outputs: Revenue ton kilometres, revenue passenger kilometres, total business income, CO_2 emissions decrease index.
Lee et al. (2015)	Malmquist-Luenberger index (output orientation, CRS)	35 international airlines, 2004–11	Inputs: Number of employees, average aircraft capacity, fuel burn, hours flown. Outputs: Tonne kilometres performed, CO_2 emissions.
Scotti and Volta (2015)	Biennial Malmquist −Luenberger index (output orientation, VRS)	18 European airlines, 2000–10	Inputs: Available seat kilometres, available freight-tonne kilometres. Outputs: Revenue passenger kilometres, total freight tonne kilometres, CO_2 emissions.
Cui and Li (2016)	Two-stage SBM-NDEA (nonoriented, VRS)	22 international airlines, 2008–12	Inputs (stage 1): Salaries, wages and benefits, fuel expenses, total assets. Intermediates: Estimated carbon dioxide. Outputs (stage 1): Revenue passenger kilometres, revenue tonne kilometres. Inputs (stage 2): Estimated carbon dioxide, abatement expense. Outputs (stage 1): Estimated carbon dioxide.
Cui, Wei, et al. (2016)	Network range-adjusted measure (nonoriented, VRS)	22 international airlines, 2008–12	Inputs (stage 1): Number of employees, aviation kerosene. Intermediates (stage 1): Available seat kilometres, available tonne kilometres. Inputs (stage 2): Available seat kilometres, available tonne kilometres, fleet size. Intermediates (stage 2): Revenue

Continued

TABLE 2.4 Airline performance studies incorporating bad outputs.—cont'd

Author(s)	Model	Sample	Inputs and outputs
			passenger kilometres, revenue tonne kilometres. Outputs (stage 2): Greenhouse gases emission (CO_2). Inputs (stage 3): Revenue passenger kilometres, revenue tonne kilometres, sales cost. Outputs: Total business income.
Li et al. (2016a)	Three-stage network SBM (nonoriented, CRS)	22 international airlines, 2008–12	Inputs (stage 1): Number of employees, aviation kerosene. Intermediates (stage 1): Available seat kilometres, available tonne kilometres. Inputs (stage 2): Available seat kilometres, available tonne kilometres, fleet size. Intermediates (stage 2): Revenue passenger kilometres, revenue tonne kilometres. Outputs (stage 2): Greenhouse gases emission (CO_2). Inputs (stage 3): Revenue passenger kilometres, revenue tonne kilometres, sales cost. Outputs: Total business income.
Cui (2017)	Network environmental production function (nonoriented, VRS)	29 international airlines, 2008–15	Inputs (stage 1): Operating expenses. Intermediates (stage 1): Available seat kilometres. Inputs (stage 2): Available seat kilometres, fleet size. Outputs (stage 2): Greenhouse gases emission (CO_2). Intermediates (stage 2): Revenue passenger kilometres. Inputs (stage 3): Revenue passenger kilometres, sales cost. Outputs (stage 3): Total revenue.
Lee et al. (2017)	Luenberger productivity index (output orientation, CRS)	34 international airlines, 2004–11	Inputs: Number of employees, average aircraft capacity, fuel burn, hours flown. Outputs: Revenue tonne kilometres, CO_2 emissions.

TABLE 2.4 Airline performance studies incorporating bad outputs.—cont'd

Author(s)	Model	Sample	Inputs and outputs
Li and Cui (2017a)	Virtual frontier network range adjusted measure (nonoriented, VRS)	22 international airlines, 2008—12	Inputs (stage 1): Number of employees, aviation kerosene. Intermediates (stage 1): Available seat kilometres, available tonne kilometres. Inputs (stage 2): Available seat kilometres, available tonne kilometres, fleet size. Intermediates (stage 2): Revenue passenger kilometres, revenue tonne kilometres. Outputs (stage 2): Greenhouse gases emission (CO_2). Inputs (stage 3): Revenue passenger kilometres, revenue tonne kilometres, sales cost. Outputs: Total business income.
Li and Cui (2017b)	Network range-adjusted environmental DEA (nonoriented, VRS)	29 international airlines, 2008—15	Inputs (stage 1): Operating expenses. Intermediates (stage 1): Available seat kilometres. Inputs (stage 2): Available seat kilometres, fleet size. Outputs (stage 2): Greenhouse gases emission (CO_2). Intermediates (stage 2): Revenue passenger kilometres. Inputs (stage 3): Revenue passenger kilometres, sales cost. Outputs (stage 3): Total revenue.
Chen et al. (2017)	Stochastic network DEA (input orientation, CRS)	13 Chinese airlines, 2006—14	Inputs: Fuel, number of planes, number of employees. Intermediates: Number of landings and take-offs Outputs: Delays (percentage), CO_2, cargo, number of passengers.
Zhang et al. (2017)	SBM (nonoriented, CRS)	18 US and Chinese airlines, 2011—14	Inputs: Number of aircraft, labour, fuel consumption. Outputs: Revenue tonne kilometres, operating revenue, CO_2 emissions.
Cui and Li (2018)	Network epsilon-based measure (nonoriented, VRS)	29 international airlines, 2008—15	Inputs (stage 1): Operating expenses. Intermediates (stage 1): Available seat kilometres.

Continued

TABLE 2.4 Airline performance studies incorporating bad outputs.—cont'd

Author(s)	Model	Sample	Inputs and outputs
			Inputs (stage 2): Available seat kilometres, fleet size. Outputs (stage 2): Greenhouse gases emission (CO_2). Intermediates (stage 2): Revenue passenger kilometres. Inputs (stage 3): Revenue passenger kilometres, sales cost. Outputs (stage 3): Total revenue.
Cui and Li (2019)	Network environmental SBM (nonoriented, VRS)	18 international airlines, 2008–14	Inputs (stage 1): Operating expenses. Intermediates (stage 1): Available seat kilometres, available tonne kilometres. Inputs (stage 2): Available seat kilometres, available tonne kilometres, fleet size. Outputs (stage 2): Greenhouse gases emission (CO_2). Intermediates (stage 2): Revenue passenger kilometres, revenue tonne kilometres. Inputs (stage 3): Revenue passenger kilometres, revenue tonne kilometres, sales cost. Outputs (stage 3): Total revenue.
Cui (2020)	Network range-adjusted measure (nonoriented, VRS)	29 international airlines, 2008–15	Inputs (stage 1): Number of employees, aviation kerosene. Intermediates (stage 1): Available seat kilometres. Inputs (stage 2): Available seat kilometres, fleet size. Intermediates (stage 2): Revenue passenger kilometres. Outputs (stage 2): Greenhouse gases emission (CO_2). Inputs (stage 3): Revenue passenger kilometres, sales cost. Outputs: Total revenue.
Xu et al. (2021)	DEA directional distance function (input orientation, CRS)	12 US airlines, 2013–16	Inputs: Employment, operating expense, fuel consumption. Outputs: Revenue passenger miles, flight delay, greenhouse gas emissions.

TABLE 2.5 Airline performance studies based on Network DEA model or DEA linked by phases.

Author(s)	Model	Sample	Inputs and outputs
Zhu (2011)	Two-stage NDEA (input orientation, CRS)	21 US airlines, 2007–08	Inputs: Cost per available seat mile, salaries per available seat mile, wages per available seat mile, benefits per available seat mile, fuel expense per available seat mile. Intermediates: Load factor, fleet size. Outputs: Revenue Passenger miles, passenger revenue.
Gramani (2012)	Two-phase DEA (input and output orientation, VRS)	34 Brazilian and US airlines, 1997 –2006	Phase 1 Inputs: Aircraft fuel, wages, salaries and benefit; cost per available seat mile. Outputs: Revenue passenger mile. Phase 2 Inputs: 1/efficiency score (inverse of the efficiency score in phase 1). Outputs: Flight revenue, flight income.
Lee and Johnson (2012)	Two-stage NDEA (output orientation, VRS)	15 US airlines, 2007–08	Inputs: Aircraft fleet size (average number of aircraft employed), fuel consumption, number of employees. Intermediates: Capacity (peak passenger output, peak freight output), scheduled demand (passenger and freight demand), available output (passenger and freight output). Outputs: Passenger and freight; realized demand.
Lu et al. (2012)	Two-stage NDEA (input orientation, CRS)	30 US airlines, 2006	Inputs: Number of employees (FTE), fuel consumed, seating capacity, flight equipment (cost), expenses (maintenance, materials, repairs), ground property and equipment (cost). Intermediates: Available seat miles (ASM), available ton miles (ATM). Outputs: Revenue passenger miles (RPMs), nonpassenger revenue (NPR).

Continued

TABLE 2.5 Airline performance studies based on Network DEA model or DEA linked by phases.—cont'd

Author(s)	Model	Sample	Inputs and outputs
Chang and Yu (2014)	Slacks-based NDEA (nonoriented, VRS)	16 low cost carriers, 2007	Inputs: Number of full-time employees, fleet size, fuel consumption. Intermediates: Available seat miles, number of destinations. Outputs: Revenue passenger miles.
Lozano and Gutiérrez (2014)	Slacks-based NDEA (nonoriented, VRS)	16 European airlines, 2007	Inputs (stage 1): Fuel cost, noncurrent assets, wages and salaries, other operating costs. Intermediates (stage 1): Available seat kilometres, available tonne kilometres. Inputs (stage 2): Selling cost. Outputs: Revenue passenger kilometres, revenue tonne kilometres.
Tavassoli et al. (2014)	SBM-NDEA (nonoriented, CRS)	11 Iranian airlines, 2010	Inputs: Number of passenger planes, labour, number of cargo planes. Intermediates: Passenger plane km, cargo plane km. Outputs: Passenger kilometres, tonne kilometres.
Li et al. (2015)	Three-stage virtual frontier network SBM (nonoriented, CRS)	22 international airlines, 2008–12	Inputs (stage 1): Number of employees, aviation kerosene. Intermediates (stage 1): Available seat kilometres, available tonne kilometres. Inputs (stage 2): Available seat kilometres, available tonne kilometres, fleet size. Intermediates (stage 2): Revenue passenger kilometres, revenue tonne kilometres. Inputs (stage 3): Revenue passenger kilometres, revenue tonne kilometres, sales costs. Outputs: Total business income.
Mallikarjun (2015)	Three-stage NDEA (nonoriented, VRS)	27 US airlines, 2012	Inputs (stage 1): Operating expenses Intermediates (stage 1): Available seat miles. Inputs (stage 2): Available seat miles, fleet size (number of aircrafts available), destinations (number of airports (destinations) served). Intermediates (stage 2): Revenue passenger miles. Inputs (stage 3): Revenue passenger miles. Outputs: Operating revenue.

Chou et al. (2016)	Meta dynamic network slack-based measure (nonoriented, VRS)	35 international airlines, 2007–09	Inputs: Labour cost, fuel cost, fleet size. Intermediates: Available seat kilometres. Carry over (stage 2): Number of accidents, net revenues (in time t-1). Outputs: Net revenues (in time t), passenger kilometers.
Cui and Li (2016)	Two-stage SBM-NDEA (nonoriented, VRS)	22 international airlines, 2008–12	Inputs (stage 1): Salaries, wages and benefits, fuel expenses, total assets. Intermediates: Estimated carbon dioxide. Outputs (stage 1): Revenue passenger kilometres, revenue tonne kilometres. Inputs (stage 2): Estimated carbon dioxide, abatement expense. Outputs (stage 1): Estimated carbon dioxide.
Cui, Li, et al. (2016)	Two-stage virtual frontier dynamic SBM	21 international airlines, 2008–12.	Inputs: Number of employees, aviation kerosene. Intermediates: Capital stock. Outputs: Revenue tonne kilometres, revenue passenger kilometres, total business income.
Cui, Wei, et al. (2016)	Network range-adjusted measure (nonoriented, VRS)	22 international airlines, 2008–12	Inputs (stage 1): Number of employees, aviation kerosene. Intermediates (stage 1): Available seat kilometres, Available tonne kilometres. Inputs (stage 2): Available seat kilometres, Available tonne kilometres, fleet size. Intermediates (stage 2): Revenue passenger kilometres, revenue tonne kilometres. Outputs (stage 2): Greenhouse gases emission (CO_2). Inputs (stage 3): Revenue passenger kilometres, revenue tonne kilometres, sales cost. Outputs (stage 3): Total business income.
Duygun et al. (2016)	Two-stage NDEA (output orientation, VRS)	87 international airlines, 2000–10	Inputs: Flight capital (number of aircrafts), labor Index (multilateral Törnqvist –Theil index constructed using the FTE of number of pilots, cabin crew, mechanics, aircraft handlers, and other labor), material quantity index (multilateral Törnqvist–Theil index, composed of the quantity of nonflight equipment, quantity of other materials, quantity of fuel, and quantity of landing services). Intermediates: Revenue ton km per load factor. Outputs: Revenue ton km.

Continued

TABLE 2.5 Airline performance studies based on Network DEA model or DEA linked by phases.—cont'd

Author(s)	Model	Sample	Inputs and outputs
Li et al. (2016b)	Virtual frontier dynamic range adjusted measure (nonoriented, VRS)	22 international airlines, 2008—12	Inputs: Number of employees, aviation kerosene. Carry over: Capital stock Outputs: Revenue passenger kilometres, revenue tonne kilometres, total business income.
Omrani and Soltanzadeh (2016)	Relational dynamic NDEA, NDEA and DDEA (input orientation, CRS)	8 Iranian airlines, 2010—12	Inputs: Number of employees. Intermediates: Available seat kilometres, available ton kilometres, number of scheduled flights. Carry over: Number of fleet's seat. Outputs: Passenger kilometres performed, passenger ton kilometres performed.
Wanke and Barros (2016)	Virtual frontier dynamic range adjusted measure DEA (nonoriented, VRS)	19 Latin American airlines, 2010—14	Inputs: Number of employees. Carry over: Number of aircraft. Outputs: Number of domestic flights, number of Latin and Caribbean flights, number of world flights.
Chen et al. (2017)	Two-stage stochastic NDEA (input orientation, CRS)	13 Chinese airlines, 2006—14	Inputs: Fuel, number of planes, number of employees. Intermediates: Number of landings and take-offs Outputs: Delays (percentage), cargo, number of passengers.
Cui (2017)	Network environmental production function (nonoriented, VRS)	29 international airlines, 2008—15	Inputs (stage 1): Operating expenses. Intermediates (stage 1): Available seat kilometres. Inputs (stage 2): Available seat kilometres, fleet size. Outputs (stage 2): Greenhouse gases emission (CO_2). Intermediates (stage 2): Revenue passenger kilometres. Inputs (stage 3): Revenue passenger kilometres, sales cost. Outputs (stage 3): Total revenue.

Cui and Li (2017)	Dynamic epsilon-based measure model (nonoriented, CRS)	19 international airlines, 2008–14	Inputs: Number of employees, aviation kerosene. Carry over: Capital stock. Outputs: Revenue passenger kilometres, revenue tonne kilometres, total revenue.
Li and Cui (2017a)	Virtual frontier network range adjusted measure (nonoriented, VRS)	22 international airlines, 2008–12	Inputs (stage 1): Number of employees, aviation kerosene. Intermediates (stage 1): Available seat kilometres, available tonne kilometres. Inputs (stage 2): Available seat kilometres, available tonne kilometres, fleet size. Intermediates (stage 2): Revenue passenger kilometres, revenue tonne kilometres. Outputs (stage 2): Greenhouse gases emission (CO_2). Inputs (stage 3): Revenue passenger kilometres, revenue tonne kilometres, sales cost. Outputs: Total business income.
Li and Cui (2017b)	Network range-adjusted Environmental DEA (nonoriented, VRS)	29 international airlines, 2008–15	Inputs (stage 1): Operating expenses. Intermediates (stage 1): Available seat kilometres. Inputs (stage 2): Available seat kilometres, fleet size. Outputs (stage 2): Greenhouse gases emission (CO_2). Intermediates (stage 2): Revenue passenger kilometres. Inputs (stage 3): Revenue passenger kilometres, sales cost. Outputs (stage 3): Total revenue.
Xu and Cui (2017)	Network epsilon-based measure and network slacks-based measure (nonoriented, VRS)	19 international airlines, 2008–14	Inputs (stage 1a): Number of employees, aviation kerosene. Inputs (stage 1b): Maintenance costs. Intermediates (stage 1): Available seat kilometres, available tonne kilometres, fleet size. Inputs (stage 2): Available seat kilometres, available tonne kilometres, fleet size, number of destinations. Intermediates (stage 2): Revenue passenger kilometres, revenue tonne kilometres. Inputs (stage 3): Revenue passenger kilometres, revenue tonne kilometres, sales costs. Outputs: Total business income.

Continued

TABLE 2.5 Airline performance studies based on Network DEA model or DEA linked by phases.—cont'd

Author(s)	Model	Sample	Inputs and outputs
Yu et al. (2017)	Two-stage DNDEA (nonoriented, VRS)	30 international airlines, 2009–12	Inputs: Number of leased aircraft, total staff wages and benefits, total fuel expenses, other operational expenses excluding labor and fuel expenses. Intermediates: Available seats kilometres, available ton kilometres. Carry over: Number of self-owned aircraft, waypoints. Outputs: Revenue passenger kilometres, freight revenue tonne kilometres.
Cui and Li (2018)	Network epsilon-based measure (nonoriented, VRS)	29 international airlines, 2021–23	Inputs (stage 1): Operating expenses. Intermediates (stage 1): Available seat kilometres. Inputs (stage 2): Available seat kilometres, fleet size. Outputs (stage 2): Greenhouse gases emission. Intermediates (stage 2): Revenue passenger kilometres. Inputs (stage 3): Revenue passenger kilometres, sales cost. Outputs (stage 3): Total revenue.
Li and Cui (2018)	Input-shared network range adjusted measure (nonoriented, VRS)	29 international airlines, 2008–15	Inputs (stage 1): Number of employees × a, aviation kerosene. Intermediates (stage 1): Available seat kilometres. Inputs (stage 2): Available seat kilometres, fleet size, number of employees × b. Intermediates (stage 2): Revenue passenger kilometres. Inputs (stage 3): Revenue passenger kilometres, sales cost, number of employees × (1−a−b). Outputs: Total revenue. * a and b are proportions of total number of employees.
Cui and Li (2019)	Network environmental SBM (nonoriented, VRS)	18 international airlines, 2008–14	Inputs (stage 1): Operating expenses. Intermediates (stage 1): Available seat kilometres, available tonne kilometres. Inputs (stage 2): Available seat kilometres, available tonne kilometres, fleet size. Outputs (stage 2): Greenhouse gases emission (CO_2). Intermediates (stage 2): Revenue passenger kilometres, revenue tonne kilometres. Inputs (stage 3): Revenue passenger kilometres, revenue tonne kilometres, sales cost. Outputs (stage 3): Total revenue.

Yu et al. (2019)	Dynamic NDEA-SBM (nonoriented, CRS)	13 Chinese and Indian airlines, 2008–15	Inputs: Number of employees, number of aircraft. Intermediates: Number of departures, flying hours. Carry over: Number of destinations served. Outputs: Revenue passenger kilometres, revenue tonne kilometres.
Cui (2020)	Network range-adjusted measure (nonoriented, VRS)	29 international airlines, 2008–15	Inputs (stage 1): Number of employees, aviation kerosene. Intermediates (stage 1): Available seat kilometres. Inputs (stage 2): Available seat kilometres, fleet size. Intermediates (stage 2): Revenue passenger kilometres. Outputs (stage 2): Greenhouse gases emission (CO_2). Inputs (stage 3): Revenue passenger kilometres, sales cost. Outputs: Total revenue.
Lin and Hong (2020)	Relational network DEA (Comprehensive oriented, CRS)	8 China and Taiwan airlines, 2003–12	Inputs: Operating expenses, number of employees. Intermediates: Revenue per available seat mile, load factor. Outputs: Revenue passenger kilometres, revenue tonne kilometres, total operating revenue.
Huang et al. (2021)	Two-stage network DEA (input orientation, CRS)	9 US airlines, 2015–19	Inputs: Cost per available seat miles. Intermediates: Available seat kilometres, available tonne kilometres. Outputs: Return on investment.

Gramani (2012) considered a VRS model because increase in inputs does not generate the same increase in outputs in both phases (i.e., nodes), which reflects a VRS model more so than a CRS model. Mallikarjun (2015) considered VRS because airlines vary in size, which is a critical factor in determining efficiency. Cui and Li (2016) considered VRS because it is more suitable for time series analysis. Xu and Cui (2017) used VRS because they noted that in practice, not all inputs or outputs behave in proportional way.

Yu et al. (2017) used VRS as the scale of operations between airline companies are different. Conversely, the following two studies used the CRS assumption. Lu et al. (2012) used the Wilcoxon Matched Pairs Test to determine between CRS and VRS model and the test supported the CRS assumption. Yu et al. (2019) assumed CRS because in the long run, airlines would adjust its capacity to move to CRS.

Gramani (2012) considered input orientation in phase I as the aim to investigate the optimization of the resources. Phase II considered output orientation because financial performance of airline companies is the objective. Lu et al. (2012) used input orientation because of the controllable aspects of a manager. Mallikarjun (2015) used nonoriented model as the objective of their study is to decrease input levels and increase output levels simultaneously. Yu et al. (2019) used a nonoriented model because airlines can effectively control their inputs while expanding outputs.

2.7 Literature on airline efficiency using other variations of DEA models

In this section, we provide literature on airline productivity and efficiency analysis that extended on the DEA model but not covered from Sections 2.2 to 2.6. Some of the studies employed free disposable hull (FDH) model whereby it relaxes the convexity assumption of the standard DEA models and thus does not require that convex combinations of every observed production plan be included in the production set (Deprins et al., 1984). Other studies used the super-efficiency DEA model as it can discriminate efficient DMU's, thus identifying inefficient DMUs which were marked as efficient under the standard DEA model. Table 2.6 presents the literature on airline performance studies based on other DEA variations that do not fall under the categories of Sections 2.2 to 2.6.

Alam and Sickles (1998, 2000) used CRS based on their test which was not rejected and also found that Cornwell et al. (1990) and Caves et al. (1984) supported the use of CRS. Kottas and Madas (2018) considered CRS because their study aimed at assessing overall efficiency.

Barros et al. (2013) assumed output orientation because the objective of the airlines is to maximize outputs. Kottas and Madas (2018) considered input orientation as airline have better ability to control their inputs than outputs.

TABLE 2.6 Airline performance studies based on other DEA variations.

Author(s)	Model	Sample	Inputs and outputs
Alam and Sickles (1998)	FDH (output orientation, CRS)	11 US airlines, 1970–90	Inputs: flight capital (number of planes), labour (pilots, flight attendants, mechanics, passenger and aircraft handlers, and other labour), energy (aircraft fuel), materials (supplies, outside services, and nonflight capital). Outputs: Quantity of revenue output (revenue ton miles) including passenger and nonscheduled revenue output.
Alam and Sickles (2000)	FDH (input orientation, CRS)	11 US airlines, 1970–90	Inputs: flight capital (number of planes), labour (pilots, flight attendants, mechanics, passenger and aircraft handlers and other labour), energy (aircraft fuel), materials (supplies, outside services and nonflight capital). Outputs: Quantity of revenue output (revenue ton miles) including passenger and nonscheduled revenue output.
Greer (2006)	Superefficiency (input orientation, CRS)	14 US airlines, 2004	Inputs: Labour, aircraft fuel, fleet-wide seating capacity. Outputs: Available seat miles.
Ray (2008)	Directional distance function and Nerlove –Luenberger superefficiency (nonoriented, VRS)	28 international airlines	Inputs: Number of employees, fuel measured (in millions of gallons), other inputs (millions of US dollar equivalent consisting of operating and maintenance expenses excluding labour and fuel expenses, capital (sum of the maximum take-off weights of all aircrafts flown multiplied by the number of days flown). Outputs: Passenger kilometres flown, freight tonne kilometres flown.
Barros et al. (2013)	DEA B-convex model (output orientation, CRS and VRS)	11 US airlines, 1998–2010	Inputs: Total cost, number of employees, fuel (in gallons). Outputs: Total revenue, revenue passenger miles, passenger load factor.

Continued

TABLE 2.6 Airline performance studies based on other DEA variations.—cont'd

Author(s)	Model	Sample	Inputs and outputs
Tavassoli et al. (2016)	Range-adjusted measure and strong complementary slackness condition DEA, DEA discriminant analysis (nonoriented, VRS)	7 Iranian airlines, 2007—11	Inputs: Number of aeroplanes, number of employees, number of flights. Outputs: Passenger plane (km), cargo plane (km).
Kottas and Madas (2018)	Superefficiency (input orientation, CRS)	30 international airlines, 2012—16	Inputs: Number of employees, total operating cost, number of operated aircraft. Outputs: Total operating revenue, revenue passenger kilometers, revenue tonne kilometres.
Wang et al. (2018)	Superefficiency SBM-DEA (input orientation, CRS)	11 ASEAN airlines, 2013—16	Inputs: Total assets, total liabilities, total equity. Outputs: Selling, general and administrative expenses, revenue.
Wang et al. (2019)	DEA window model and Grey model (Orientation unspecified, returns to scale unspecified)	17 Asian airlines, 2012—16	Inputs: Total assets, number of operating aeroplanes, operating expenses. Outputs: Available seat kilometres, revenue passenger kilometres.
Hermoso et al. (2019)	Net-DEA, SM-DEA (output orientation, CRS and VRS)	43 European airlines, 2014	Inputs (for Net-DEA): Number of employees, total assets, number of destinations, degree centrality, eigen centrality. Outputs (for Net-DEA): Sales, number of passengers. Inputs (for SM-DEA): Number of employees, total assets, number of destinations, tweets per day, publication per day, number of videos. Outputs (for SM-DEA): Sales, number of passengers, likes in twitter, likes in facebook, views.

TABLE 2.6 Airline performance studies based on other DEA variations.—cont'd

Author(s)	Model	Sample	Inputs and outputs
Lin and Hong (2020)	Metafrontier DEA (input and output orientation, CRS)	8 China and Taiwan airlines, 2003–12	Inputs: Operating expenses, number of employees. Intermediates: Revenue per available seat mile, load factor. Outputs: Revenue passenger kilometres, revenue tonne kilometres, total operating revenue.
Ngo and Tsui (2022)	DEA-Stochastic (input orientation, CRS)	14 Asia –Pacific airlines, 2008–15	Inputs: Available seat kilometres, available tonne kilometres, operating expenses. Outputs: Revenue passenger kilometres, revenue tonne kilometres (RTK), operating revenues.

2.8 Literature on airline efficiency incorporating second-stage regression analysis

In this section, we provide the literature on airline efficiency and productivity analysis incorporating a second-stage regression. From the literature, two time periods can be discerned—pre-2007 and post-2007. For pre-2007, the main regression models were ordinary least squares (OLS) and Tobit. For post-2007, the main regression models were Simar and Wilson's (SW) bootstrap truncated regression, Tobit regression and generalized least squares (GLS).

The segmentation of the two time periods is attributed to the study of Simar and Wilson (2007) whereby they argued that directly employing the second-stage regression on the DEA efficiency scores does not provide any coherent data-generating process (DGP), which cast doubts on what is being estimated. Simar and Wilson (2007) also argued that regressing DEA efficiency scores on covariates (i.e., environmental variables) suffer from the problem of serial correlation. Hence, direct regression analysis is invalid, inconsistent and biased owing to the dependency of the efficiency scores. However, Banker et al. (2019) contested Simar and Wilson's bootstrap truncated regression by performing Monte Carlo simulations and found that both the OLS method and to a lesser extent the Tobit method outperformed the SW's method.

The purpose of this section is not to critically evaluate the pros and cons of each regression model or provide any recommendations. The aim is to provide

TABLE 2.7 Airline performance studies incorporating second-stage regression.

Author(s)	Model	Sample	Environmental variables
Schefczyk (1993)	OLS	15 international airlines, 1990	Aircraft flight length, nonflight assets as a percentage of available ton km, passenger revenues as a percentage of total revenues, international passenger revenue km as a percentage of total passenger revenue km, percentage of state ownership in the airline.
Banker and Johnston (1994)	OLS	12 US airlines, 1981–85 (quarterly data)	Model 1 (input usage inefficiency): Percentages of flights through competitive and dominated hubs, the average load factor, a surrogate measure of service quality. Model 2 (revenue generating inefficiency): Percentages of flights through competitive and dominated hubs, the average load factor, a surrogate measure of service quality. Model 3 (use of flight equipment inefficiency): Hub concentration variables, load factor, aircraft utilization rate. Model 4 (use of flight crew labour inefficiency): RPM in wide-bodied and fuel-efficient aircraft, load factors, quality. Model 5 (fuel inefficiency): RPM in wide-bodied and fuel-efficient aircraft, load factor.
Alam and Sickles (2000)	Tobit	11 US airlines, 1970–90	Aircraft stage length, load factor, average size of the carrier's aircraft, the percentage of a carrier's fleet which is jet.
Fethi et al. (2000)	Tobit	17 European airlines, 1991–95	Overall load factor, state ownership, European flights, international flights, route network density, concentration (dummy variable), subsidies (dummy variable), year (dummy variable).

TABLE 2.7 Airline performance studies incorporating second-stage regression.—cont'd

Author(s)	Model	Sample	Environmental variables
Desli and Ray (2004)	OLS	21 US airlines, 1984	Stage length (average distance flown between takeoff and landing), passenger load factor, the number of points served.
Scheraga (2004)	Tobit	38 international airlines, 1995 and 2000	Aircraft flight length, nonflight assets as a percentage of available ton km, passenger revenues as a percentage of total revenues, international passenger revenue km as a percentage of total passenger revenue km, percentage of state ownership in the airline, available ton km, percentage of load factor (ton km).
Greer (2006)	Log-linear regression	14 US airlines, 2004	Aircraft stage length, discount or legacy airline (dummy variable).
Scheraga (2006)	Tobit	34 international airlines, 2003 and 2004	Average flight length, passenger revenues as a percentage of total revenues, scheduled service revenues as a percentage of total revenues, international passenger revenue kilometers as a percentage of total passenger revenue kilometers, average load factor, expenditures on ticketing sales and promotion per revenue passenger.
Barbot et al. (2008)	OLS	49 international airlines, 2005	Employees per million ASK, block hours per day, fuel consumption per million ASKs, airlines' size (in ASKs), low-cost carrier (dummy variable).
Barros and Peypoch (2009)	Simar and Wilson bootstrap truncated regression	29 international airlines, 2000–05	Local market (population), managerial practice (low cost vs national airline), Star alliance, One World, Sky Team, trend, square trend.
Bhadra (2009)	Tobit regression	13 US airlines, 1985–2006	Load factor, revenue passenger miles, yield, number of departures, number of passengers, aircraft days, block hours, operating ratio, fuel cost per gallons, pay per FTEs, terror (dummy variable).

Continued

TABLE 2.7 Airline performance studies incorporating second-stage regression.—cont'd

Author(s)	Model	Sample	Environmental variables
Assaf (2011)	Tobit regression	18 UK airlines, 2004–07	Average load factor, average stage length, airline size (dummy variable and based on the percentage share of each airline in the total kilometres performed), low-cost carrier (dummy variable).
Assaf and Josiassen (2011)	Simar and Wilson bootstrap truncated regression	15 UK airlines, 2002–07	Load factor, airline size (dummy variable).
Merkert and Hensher (2011)	Tobit regression with bootstrapping	58 international airlines, 2007/08 and 2008/09	Available tonne kilometres, average number of seats per aircraft, average stage length, fleet age, number of different aircraft families.
Wang et al. (2011)	Simar and Wilson bootstrap truncated regression	30 international airlines, 2006	Board size (number of directors), committees (number of committees servicing the board), annual number of board meetings, nonexecutive director, CEO duality (dummy variable).
Lu et al. (2012)	Simar and Wilson bootstrap truncated regression	30 US airlines, 2006	Board size (the number of directors on the board including executive and nonexecutive directors), committees (the number of committees established by the board, meetings (the annual number of board meetings), nonexecutive director (the number of independent nonexecutive directors on the board), CEO duality (dummy variable for airlines, which equals 1 if the CEO is also chairman of the board, and 0 otherwise), directors' age (average age of the board directors), executive officers ownership (the percent of the firm's outstanding shares owned by the executive officers),

TABLE 2.7 Airline performance studies incorporating second-stage regression.—cont'd

Author(s)	Model	Sample	Environmental variables
			average age of aircraft, average aircraft size (average number of seats on aircraft), average stage length (average distance flown on each segment of every route), dummy international (dummy variable for airlines, which equals 1 if the airline has international flights, and 0 otherwise), dummy low-cost carrier (dummy variable for airlines, which equals 1 if the airline is a low-cost carrier, and 0 otherwise).
Wu et al. (2013)	Simar and Wilson bootstrap truncated regression	12 international airlines, 2006—10	INTER (the ratio of international RPK to total RPK), CARGO (the percentage of cargo revenue to the total operating revenue), squared value of CARGO, SALARIES (the total staff costs divided by the number of full time employees), NATIONALITY (dummy variable), LOGPOPU (log transformation of respective population), LOAD (percentage of RTK to available tonne kilometers), FUEL (ratio of total fuel costs to ATK).
Lee and Worthington (2014)	Simar and Wilson bootstrap truncated regression	42 international airlines, 2006	Ownership (dummy variable), low-cost carriers (dummy variable), number of departures, and weight load factor.
Cui, Li, et al. (2016)	Tobit regression model with smoothing homogeneous bootstrap	21 international airlines, 2008—12.	Proportion of aircrafts with the largest number to the total aircrafts, per capita GDP of the country or region that the airline headquarters is located, average service age of fleet size, number of destinations, average haul distance.

Continued

TABLE 2.7 Airline performance studies incorporating second-stage regression.—cont'd

Author(s)	Model	Sample	Environmental variables
Sarangal and Nagpal (2016)	GLS and Tobit	59 Indian airlines, 2005—06 to 2011—12	Yield (revenue per RPK), average daily revenue hours per aircraft, average stage distance flown per aircraft departure, percentage of international RPK, low-cost carriers vs full service airlines, operating expense per RPK.
Wanke and Barros (2016)	Simplex regression	19 Latin American airlines, 2010—14	Cargo operation (dummy variable), public ownership (dummy variable), percentage of large aircrafts, percentage of small aircrafts, low-cost carrier (dummy variable), trend, trend 2.
Chen et al. (2017)	Bootstrapped Tobit, Beta, Simplex regression	13 Chinese airlines, 2006—14	Age, fleet mix (percentage of Boeing aircraft), stock market governance (listed in stock market), ownership type (whether public or private), whether it has undergone merger and acquisition process, network span (whether the airline flies both international and domestic flights).
Choi (2017)	Simar and Wilson bootstrap truncated regression	14 US airlines, 2006 to 2015	Fuel expense, passenger revenue, full-time employee equivalents, total operating revenue.
Lee et al. (2017)	GLS	34 international airlines, 2004—11	Membership (dummy variable), low-cost carriers (dummy variable), ownership (dummy variable), weight load factor, average fleet age.
Yu et al. (2017)	Simar and Wilson bootstrap truncated regression	30 international airlines, 2009—12	Alliance (SkyTeam dummy variable), alliance (OneWorld dummy variable), Alliance (Star Alliance dummy variable), total assets, GDP, time effect (2009 dummy variable), time effect (2010 dummy variable), time effect (2011 dummy variable).

TABLE 2.7 Airline performance studies incorporating second-stage regression.—cont'd

Author(s)	Model	Sample	Environmental variables
Zhang et al. (2017)	Tobit regression	18 US and Chinese airlines, 2011–14	Fleet age, passengers per flight, freight traffic, haul distance, ownership, market share.
Mhlanga et al. (2018)	GLS and Tobit regression	7 South African airlines, 2012–15	Yield (revenue per revenue passenger kilometre), seat load factor (%), average daily revenue hours per aircraft, aircraft families, average aircraft size (seats), average daily hours blocked per aircraft, low-cost carrier (dummy variable).
Yu et al. (2019)	Tobit regression	13 Chinese and Indian airlines, 2008–15	Low-cost carrier (dummy variable), ownership (dummy variable), average route-level HHI, share of international revenue passenger kilometres, average stage length per trip, length of high-speed rail.
Huang et al. (2021)	Simar and Wilson bootstrap truncated regression	9 US airlines, 2015–19	Average stage length flown, total assets, fleet size, full-time employee equivalents.
Ngo and Tsui (2022)	Tobit regression	14 Asia –Pacific airlines, 2008–15	National carrier (dummy variable), alliance (dummy variable), listed in stock market (dummy variable), total assets, global financial crisis (dummy variable), MH_accidents (dummy variable), GDP per capita, international tourist arrivals, trade volume, fuel prices.

a range of regression models and inform the reader of the models that have been used in airline efficiency and productivity analysis studies. Hence, researchers will need to decide among the various second-stage regression models appropriate for their research study.

Table 2.7 provides the literature of airline performance studies that employed a second-stage regression. As observed, since the paper of Barros and Peypoch (2009), the number of airline studies that employed Simar and

Wilson (2007) bootstrap truncated regression increased considerably. Some studies such as Merkert and Hensher (2011), Cui, Li, et al. (2016) and Chen et al. (2017) continued the use of Tobit but applied a smoothing homogeneous bootstrap approach to satisfy the DGP and avoid the serial correlation problem of conventional two-stage DEA studies.

2.9 Conclusion

In this chapter, we examined the various DEA models in the literature on airline efficiency and productivity. We observe that the focus and motivation of a study plays an important part in determining the orientation model. In terms of the choice of CRS or VRS, this was either determined by the assumptions relating to the state of the airline industry or determined via hypothesis testing. Measuring airline performance using DEA models may also be inadequate in providing the underlying cause of an airline inefficiency. Hence, a growing number of studies have incorporated a second-stage regression analysis to determine sources of inefficiency.

The airline performance literature also provides an interesting aspect regarding the standard DEA model and the measurement of airline efficiency and productivity. Since the earliest study in 1991, the past 30 years on airline performance studies have shown a shift toward the use of NDEA incorporating bad outputs. This is evident that the recent airline performance studies are focusing more on measuring the impact airlines have on the environment, a phenomenon similar in DEA studies in agriculture and the energy sector.

References

Alam, I.M.S., Sickles, R.C., 1998. The relationship between stock market returns and technical efficiency innovations: evidence from the US airline industry. Journal of Productivity Analysis 9, 35–51.

Alam, I.M.S., Sickles, R.C., 2000. Time series analysis of deregulatory dynamics and technical efficiency: the case of the U.S. airline industry. International Economic Review 41, 203–218.

Assaf, A., 2011. A fresh look at the productivity and efficiency changes of UK airlines. Applied Economics 43, 2165–2175.

Assaf, A.G., Josiassen, A., 2011. The operational performance of UK airlines: 2002–2007. Journal of Economics Studies 38, 5–16.

Banker, R., Natarajan, R., Zhang, D., 2019. Two-stage estimation of the impact of contextual variables in stochastic frontier production function models using data envelopment analysis: second stage OLS versus bootstrap approaches. European Journal of Operational Research 278, 368–384.

Banker, R.D., Charnes, A., Cooper, W.W., 1984. Some models for estimating technical and scale inefficiencies in data envelopment analysis. Management Science 30, 1078–1092.

Banker, R.D., Johnston, H.H., 1994. Evaluating the impacts of operating strategies on efficiency in the U.S. airline industry. In: Charnes, A., Cooper, W.W., Lewin, A.Y., Seiford, L.M. (Eds.),

Data Envelopment Analysis: Theory, Methodology, and Applications. Kluwer Academic, Boston, pp. 97–128.

Barbot, C., Costa, Á., Sochirca, E., 2008. Airlines performance in the new market context: a comparative productivity and efficiency analysis. Journal of Air Transport Management 14, 270–274.

Barros, C.P., Couto, E., 2013. Productivity analysis of European airlines, 2000–2011. Journal of Air Transport Management 31, 11–13.

Barros, C.P., Liang, Q.B., Peypoch, N., 2013. The technical efficiency of US airlines. Transportation Research Part A 50, 139–148.

Barros, C.P., Peypoch, N., 2009. An evaluation of European airlines' operational performance. International Journal of Production Economics 122, 525–533.

Bhadra, D., 2009. Race to the bottom or swimming upstream: performance analysis of US airlines. Journal of Air Transport Management 15, 227–235.

Capobianco, H.M.P., Fernandez, E., 2004. Capital structure in the world airline industry. Transportation Research Part A 38, 421–434.

Caves, D.W., Christensen, L.R., Tretheway, M.W., 1984. Economies of density versus economies of scale: why trunk and local service airline costs differ. Rand Journal of Economics 15, 471–489.

Chan, P.S., Sueyoshi, T., 1991. Environmental change, competition, strategy, structure and firm performance. An application of data envelopment analysis in the airline industry. International Journal of Systems Science 22, 1625–1636.

Chang, Y.-C., Yu, M.-M., 2014. Measuring production and consumption efficiencies using the slack-based measure network data envelopment analysis approach: the case of low-cost carriers. Journal of Advanced Transportation 48, 15–31.

Chang, Y.-T., Park, H.-S., Jeong, J.-B., Lee, J.-W., 2014. Evaluating economic and environmental efficiency of global airlines: a SBM-DEA approach. Transportation Research Part D 27, 46–50.

Charnes, A., Cooper, W.W., Rhodes, E., 1978. Measuring the efficiency of decision making units. European Journal of Operational Research 2, 429–444.

Charnes, A., Gallegos, A., Li, H., 1996. Robustly efficient parametric frontiers via multiplicative DEA for domestic and international operations of the Latin American airline industry. European Journal of Operational Research 88, 525–536.

Chen, Z., Wanke, P., Antunes, J.J.M., Zhang, N., 2017. Chinese airline efficiency under CO_2 emissions and flight delays: a stochastic network DEA model. Energy Economics 68, 89–108.

Choi, K., 2017. Multi-period efficiency and productivity changes in US domestic airlines. Journal of Air Transport Management 59, 18–25.

Choi, K., Lee, D., Olson, D.L., 2015. Service quality and productivity in the US airline industry: a service quality-adjusted DEA model. Service Business 9, 137–160.

Chou, H.-W., Lee, C.-Y., Chen, H.-K., Tsai, M.-Y., 2016. Evaluating airlines with slack-based measures and metafrontiers. Journal of Advanced Transportation 50, 1061–1089.

Chow, C.K.W., 2010. Measuring the productivity changes of Chinese airlines: the impact of the entries of non-state-owned carriers. Journal of Air Transport Management 16, 320–324.

Coelli, T., Grifell-Tatje, E., Perelman, S., 2002. Capacity utilization and profitability: a decomposition of short-run profit efficiency. International Journal of Production Economics 79, 261–278.

Cornwell, C., Schmidt, P., Sickles, R., 1990. Production frontiers with cross-sectional and time-series variation in efficiency levels. Journal of Econometrics 46, 185–200.

Cui, Q., 2017. Will airlines' pollution abatement costs be affected by CNG2020 strategy? an analysis through a network environmental production function. Transportation Research Part D 57, 141–154.

Cui, Q., 2020. Airline energy efficiency measures using a network range-adjusted measure with unified natural and managerial disposability. Energy Efficiency 13, 1195–1211.

Cui, Q., Li, Y., 2015. Evaluating energy efficiency for airlines: an application of VFB-DEA. Journal of Air Transport Management 44–45, 34–41.

Cui, Q., Li, Y., 2016. Airline energy efficiency measures considering carbon abatement: a new strategic framework. Transportation Research Part D 49, 246–258.

Cui, Q., Li, Y., 2017. Airline efficiency measures using a dynamic epsilon-based measure model. Transportation Research Part A 100, 121–134.

Cui, Q., Li, Y., 2018. CNG2020 strategy and airline efficiency: a network epsilon-based measure with managerial disposability. International Journal of Sustainable Transportation 12, 313–323.

Cui, Q., Li, Y., 2019. Investigating the impacts of the EU ETS emission rights on airline environmental efficiency via a network environmental SBM model. Journal of Environmental Planning and Management 62, 1465–1488.

Cui, Q., Li, Y., Yu, C.-L., Wei, Y.M., 2016. Evaluating energy efficiency for airlines: an application of virtual frontier dynamic slacks based measure. Energy 113, 1231–1240.

Cui, Q., Wei, Y.M., Yu, C.L., Li, Y., 2016. Measuring the energy efficiency for airlines under the pressure of being included into the EU ETS. Journal of Advanced Transportation 50, 1630–1649.

Deprins, D., Simar, L., Tulkens, H., 1984. Measuring labor inefficiency in post offices. In: Marchand, M., Pestieau, P., Tulkens, H. (Eds.), The Performance of Public Enterprises: Concepts and Measurements. North-Holland, Amsterdam, pp. 243–267.

Desli, E., Ray, S.D., 2004. A bootstrap-regression procedure to capture unit specific effects in data envelopment analysis. Indian Economic Review 39, 89–110.

Distexhe, V., Perelman, S., 1994. Technical efficiency and productivity growth in an era of deregulation: the case of airlines. Swiss Journal of Economics and Statistics 130, 669–689.

Duygun, M., Prior, D., Shaban, M., Tortosa-Ausina, E., 2016. Disentangling the European airlines efficiency puzzle: a network data envelopment analysis approach. Omega 60, 2–14.

Färe, R., 1991. Measuring Farrell efficiency for a firm with intermediate inputs. Academia Economic Papers 19, 329–340.

Färe, R., Grosskopf, S., 1996. Intertemporal Production Frontiers: With Dynamic DEA. Kluwer Academic Publishers, Boston.

Färe, R., Grosskopf, S., 2000. Network DEA. Socio-Economic Planning Sciences 34, 35–49.

Fethi, M.D., Jackson, P., Weyman-Jones, T.G., 2000. Measuring the efficiency of European airlines: an application of DEA and Tobit analysis, EPRU discussion papers. available at: http://hdl.handle.net/2381/370.

Good, D.H., Röller, L.H., Sickles, R.C., 1995. Airline efficiency differences between Europe and the US: implications for the pace of EC integration and domestic regulation. European Journal of Operational Research 80, 508–518.

Gramani, M.C.N., 2012. Efficiency decomposition approach: a cross-country airline analysis. Expert Systems with Applications 39, 5815–5819.

Greer, M.R., 2006. Are the discount carriers actually more efficient than the legacy carriers? A data envelopment analysis. International Journal of Transport Economics 33, 37–55.

Greer, M.R., 2008. Nothing focuses the mind on the productivity quite like the fear of liquidation: changes in airline productivity in the United States, 2000–2004. Transportation Research Part A 42, 414–426.

Hermoso, R., Latorre, M.P., Martinez-Nuñez, M., 2019. Multivariate data envelopment analysis to measure airline efficiency in European airspace: a network-based approach. Applied Sciences 9, 5312. https://doi.org/10.3390/app9245312.

Hong, S., Zhang, A., 2010. An efficiency study of airlines and air cargo/passenger divisions: a DEA approach. World Review of Intermodal Transportation Research 3, 137–149.

Hu, J.-L., Li, Y., Tung, H.-J., 2017. Operational efficiency of ASEAN airlines: based on DEA and bootstrapping approaches. Management Decision 55, 957–986.

Huang, C.C., Hsu, C.C., Collar, E., 2021. An evaluation of the operational performance and profitability of the U.S. airlines. International Journal of Global Business and Competitiveness 16, 73–85.

Jain, R.K., Natarajan, R., 2015. A DEA study of airlines in India. Asia Pacific Management Review 20, 285–292.

Kottas, A.T., Madas, M.A., 2018. Comparative efficiency analysis of major international airlines using data envelopment analysis: exploring effects of alliance membership and other operational efficiency determinants. Journal of Air Transport Management 70, 1–17.

Lee, B.L., Wilson, C., Pasurka, C.A., 2015. The good, the bad, and the efficient: productivity, efficiency, and technical change in the airline industry, 2004–11. Journal of Transport Economics and Policy 49, 338–354.

Lee, B.L., Wilson, C., Pasurka, C.A., Fujii, H., Managi, S., 2017. Sources of airline productivity from carbon emissions: an analysis of operational performance under good and bad outputs. Journal of Productivity Analysis 47, 223–246.

Lee, B.L., Worthington, A.C., 2014. Technical efficiency of mainstream airlines and low-cost carriers: New evidence using bootstrap data envelopment analysis truncated regression. Journal of Air Transport Management 38, 15–20.

Lee, C.-Y., Johnson, A.L., 2012. Two-dimensional efficiency decomposition to measure the demand effect in productivity analysis. European Journal of Operational Research 216, 584–593.

Li, Y., Cui, Q., 2017a. Airline energy efficiency measures using the virtual frontier network RAM with weak disposability. Transportation Planning and Technology 40, 479–504.

Li, Y., Cui, Q., 2017b. Carbon neutral growth from 2020 strategy and airline environmental inefficiency: a network range adjusted environmental data envelopment analysis. Applied Energy 199, 13–24.

Li, Y., Cui, Q., 2018. Airline efficiency with optimal employee allocation: an input-shared network range adjusted measure. Journal of Air Transport Management 73, 150–162.

Li, Y., Wang, Y.-Z., Cui, Q., 2015. Evaluating airline efficiency: an application of virtual frontier network SBM. Transportation Research Part E 81, 1–17.

Li, Y., Wang, Y.-Z., Cui, Q., 2016a. Has airline efficiency affected by the inclusion of aviation into European Union emission trading scheme? evidences from 22 airlines during 2008–2012. Energy 96, 8–22.

Li, Y., Wang, Y.Z., Cui, Q., 2016b. Energy efficiency measures for airlines: an application of virtual frontier dynamic range adjusted measure. Journal of Renewable and Sustainable Energy 8, 207–232.

Lin, Y.-H., Hong, C.-F., 2020. Efficiency and effectiveness of airline companies in Taiwan and Mainland China. Asia Pacific Management Review 25, 13–22.

Lozano, S., Gutiérrez, E., 2014. A slacks-based network DEA efficiency analysis of European airlines. Transportation Planning and Technology 37, 623–637.

Lu, W.M., Wang, W.K., Hung, S.W., Lu, E.T., 2012. The effects of corporate governance on airline performance: production and marketing efficiency perspectives. Transportation Research Part E 48, 529–544.

Mallikarjun, S., 2015. Efficiency of US airlines: a strategic operating model. Journal of Air Transport Management 43, 46—56.

Merkert, R., Hensher, D.A., 2011. The impact of strategic management and fleet planning on airline efficiency—A random effects Tobit model based on DEA efficiency scores. Transportation Research Part A 45, 686—695.

Merkert, R., Morrell, P.S., 2012. Mergers and acquisitions in aviation—management and economic perspectives on the size of airlines. Transportation Research Part E 48, 853—862.

Mhlanga, O., Steyn, J., Spencer, J., 2018. The airline industry in South Africa: drivers of operational efficiency and impacts. Tourism Review 73, 389—400.

Min, H., Joo, S.J., 2016. A comparative performance analysis of airline strategic alliances using data envelopment analysis. Journal of Air Transport Management 52, 99—110.

Ngo, T., Tsui, K.W.H., 2022. Estimating the confidence intervals for DEA efficiency scores of Asia-Pacific airlines. Operational Research 22, 3411—3434.

Nolan, J., Ritchie, P., Rowcroft, J., 2014. International mergers and acquisitions in the airline industry. In: The Economics of International Airline Transport. J. Peoples, vol. 2014. Emerald Publishing Limited, Bingley, England, pp. 127—150.

Omrani, H., Soltanzadeh, E., 2016. Dynamic DEA models with network structure: an application for Iranian airlines. Journal of Air Transport Management 57, 52—61.

Ouellette, P., Petit, P., Tessier-Parent, L.P., Vigeant, S., 2010. Introducing regulation in the measurement of efficiency, with an application to the Canadian air carriers industry. European Journal of Operational Research 200, 216—226.

Pires, H.M., Fernandes, E., 2012. Malmquist financial efficiency analysis for airlines. Transportation Research Part E 48, 1049—1055.

Pitt, I.L., Norsworthy, J.R., 1999. Economics of the US. Commercial Airline Industry: Productivity, Technology, and Deregulation. Kluwer Academic Publishers, Boston.

Ray, S.C., 2008. The directional distance function and measurement of super-efficiency: an application to airlines data. Journal of the Operational Research Society 59, 788—797.

Ray, S.C., Mukherjee, K., 1996. Decomposition of the Fisher ideal index of productivity: a nonparametric dual analysis of US airlines data. Economic Journal 106, 1659—1678.

Sakthidharan, V., Sivaraman, S., 2018. Impact of operating cost components on airline efficiency in India: a DEA approach. Asia Pacific Management Review 23, 258—267.

Saranga, H., Nagpal, R., 2016. Drivers of operational efficiency and its impact on market performance in the Indian Airline industry. Journal of Air Transport Management 53, 165—176.

Schefczyk, M., 1993. Operational performance of airlines: an extension of traditional measurement paradigms. Strategic Management Journal 14, 301—317.

Scheraga, C.A., 2004. Operational efficiency versus financial mobility in the global airline industry: a data envelopment and Tobit analysis. Transportation Research Part A 38, 384—404.

Scheraga, C.A., 2006. The operational impacts of governmental restructuring of the airline industry in China. Journal of the Transportation Research Forum 45, 71—86.

Scotti, D., Volta, N., 2015. An empirical assessment of the CO_2-sensitive productivity of European airlines from 2000 to 2010. Transportation Research Part D 37, 137—149.

Sengupta, J.K., 1998. Testing allocative efficiency by data envelopment analysis. Applied Economics Letters 5, 689—692.

Sickles, R.C., Good, D.H., Getachew, L., 2002. Specification of distance functions using semi-and nonparametric methods with an application to the dynamic performance of eastern and western European air carriers. Journal of Productivity Analysis 17, 133—155.

Simar, L., Wilson, P.W., 2007. Estimation and inference in two-stage, semi parametric models of production processes. Journal of Econometrics 136, 31−64.

Siregar, D.D., Norsworthy, J.R., 2001. Pre- and Post-deregulation Financial Performance and Efficiency in US Airlines. IEMC'01 Proceedings. Change Management and the New Industrial Revolution, pp. 421−429. Cat. No.01CH37286.

Tavassoli, M., Badizadeh, T., Saen, R.F., 2016. Performance assessment of airlines using range-adjusted measure, strong complementary slackness condition, and discriminant analysis. Journal of Air Transport Management 54, 42−46.

Tavassoli, M., Faramarzi, G.R., Saen, R.F., 2014. Efficiency and effectiveness in airline performance using a SBM-NDEA model in the presence of shared input. Journal of Air Transport Management 34, 146−153.

Tofallis, C., 1997. Input efficiency profiling: an application to airlines. Computers & Operations Research 24, 253−258.

Wang, C.N., Dang, D.C., Van Thanh, N.V., Tran, T.T., 2018. Grey model and DEA to form virtual strategic alliance: the application for ASEAN aviation industry. International Journal of Advanced and Applied Sciences 5, 25−34.

Wang, C.-N., Tsai, T.-T., Hsu, H.-P., Nguyen, L.-H., 2019. Performance evaluation of major Asian airline companies using DEA window model and grey theory. Sustainability 11, 2701.

Wang, W.K., Lu, W.M., Tsai, C.J., 2011. The relationship between airline performance and corporate governance amongst US listed companies. Journal of Air Transport Management 17, 148−152.

Wanke, P., Barros, C.P., 2016. Efficiency in Latin American airlines: a two-stage approach combining virtual. Frontier dynamic DEA and simplex regression. Journal of Air Transport Management 54, 93−103.

White, L.J., 1979. Economies of scale and the question of natural monopoly in the airline industry. The Journal of Air Law and Commerce 44, 545−573.

Wu, W.Y., Liao, Y.K., 2014. A balanced scorecard envelopment approach to assess airlines' performance. Industrial Management & Data Systems 114, 123−143.

Wu, Y., He, C., Cao, X., 2013. The impact of environmental variables on the efficiency of Chinese and other non-Chinese airlines. Journal of Air Transport Management 29, 35−38.

Xu, X., Cui, Q., 2017. Evaluating airline energy efficiency: an integrated approach with network epsilon-based measure and network slacks-based measure. Energy 122, 274−286.

Xu, Y., Park, Y.S., Park, J.D., Cho, W., 2021. Evaluating the environmental efficiency of the U.S. airline industry using a directional distance function DEA approach. Journal of Management Analytics 8, 1−18.

Yang, C., Wang, T.-P., 2016. Productivity comparison of European airlines: bootstrapping Malmquist indices. Applied Economics 48, 5106−5116.

Yu, H., Zhang, Y., Zhang, A., Wang, K., Cui, Q., 2019. A comparative study of airline efficiency in China and India: a dynamic network DEA approach. Research in Transportation Economics 76, 100746.

Yu, M.M., Chen, L.H., Hui, C., 2017. The effects of alliances and size on airlines' dynamic operational performance. Transportation Research Part A 106, 197−214.

Zhang, J., Fang, H., Wang, H., Jia, M., Wu, J., Fang, S., 2017. Energy efficiency of airlines and its influencing factors: a comparison between China and the United States. Resources, Conservation and Recycling 125, 1−8.

Zhu, J., 2011. Airlines performance via two-stage network DEA approach. Journal of Centrum Cathedra 4, 260−269.

Chapter 3

Measuring airline performance: standard DEA

3.1 Introduction

From this chapter onwards, we will be measuring the performance of airlines using DEA and variations of it. In this chapter, we first start by measuring airline efficiency based on CRS and VRS input-orientated DEA models, followed up by cost and revenue efficiencies. For each DEA approach, we provide learners on how to interpret the efficiency scores derived from DEA and other results such as proportionate (radial) movements and slack movements. For purposes of application and academic interpretation and consistency limited by data availability, we consider 16 international airlines throughout the book. The 16 airlines are presented in Table A.1.

The number of airlines, variables and years are dependent on the DEA models we use in subsequent chapters. As for now, any cross-sectional analysis will only consider the year 2009. Hence the cross-sectional data refer only to the year 2009. Our primary source of data is drawn from the World Air Transport Statistics eDataSeries—61st Edition (2017) database (henceforth WATS (2017)). This source provides time series data that reflect airline operations and financials, which satisfy not only the current chapter's need for analyzing airline performance, but also the latter chapters. Secondary sources of data are occasionally used such as those from RDC Aviation and airline annual reports.

The chapter is divided into seven sections. Following the introduction in Section 3.1, we consider some data issues related to airline data and data requirements to satisfy DEA in Section 3.2. Section 3.3 introduces the standard DEA models based on CRS and VRS, and the cost and revenue efficiency models. Section 3.4 introduces the R package. Section 3.5 presents the R script for DEA, results, and interpretation of results base on the models described in Section 3.2. Section 3.6 provides some approaches to test the reliability of results. This chapter concludes with some brief comments in Section 3.7.

Productivity and Efficiency Measurement of Airlines. https://doi.org/10.1016/B978-0-12-812696-7.00004-0

3.2 Data issues

"Garbage in, garbage out"

As in any empirical model, data is a major issue. It is essential to define the appropriate inputs and outputs that satisfy the proposed production model. In the literature on airline performance, most studies used revenue passenger kilometres (RPK) or tonne kilometres performed (TKP) as output. These include Schmidt and Sickles (1984), Barla and Perelman (1989), Cornwell et al. (1990), Schefczyk (1993), Baltagi et al. (1995), Oum and Yu (1995), Coelli et al. (1999), Sengupta (1999), Oum et al. (2005), Scheraga (2004), Vasigh and Fleming (2005), Barbot et al. (2008), Greer (2008), Bhadra (2009), and Assaf and Josiassen (2012). Lee and Worthington (2014) used available tonne-kilometres (ATK), whereas Bhadra (2009) used available seat kilo-metres (ASK). Panayotis et al. (2009) used total annual passenger kilometres. Some studies have also considered financial indicators such as 'earnings before interest and taxes' (EBIT) or revenues as outputs. These include Barros and Peypoch (2009) and Assaf and Josiassen (2011).

Because of the myriad of studies that use a range of outputs, it is important to consider the business objective of an airline to determine the production function. If the major objective of an airline is to maximize revenue, which is a reasonable assumption, then we need to correspond the revenue with appro-priate inputs. In the case of airlines, there are a range of inputs one could consider, such as the number of pilots, cabin crew, maintenance and overhaul staff, operating expenses, marketing and promotions and several others. However, we cannot assume that all these inputs produce revenue. For example, it would be meaningless to say that maintenance and overhaul staff, pilots, cabin crew and operating expenses produce revenue. There is therefore a need for a proper production function to accurately capture inputs that produce outputs.

Studies such as Lee and Johnson (2011), Zhu (2011), Lu et al. (2012) and Lozano and Gutiérrez (2014) provide a meaningful depiction of the airline production model, via a two-stage production model instead of a black box approach. It is therefore possible to develop a number of production models that represent different aspects of the airline production model. For example, we know that maximizing revenue is a major objective and outcome of air-lines, which is directly related to the number of passengers and freight tonnes carried, and ticketing, sales and promotion staff. Another production model is the service quality of airlines. Outputs for this would include flight service quality and percentage of on-time arrivals. Inputs for such a model would comprise pilots, cabin crew and hours flown. Airline production model could also measure the performance of airlines in terms of TKP. For this model, inputs would comprise pilots, cabin crew, fuel burned and number and configuration of aircraft.

From the aforementioned description, we can see that the airline production model comprise different productive activities, which can be expressed as a supply chain model. However, unlike Zhu (2011), Lee and Johnson (2012), Lu et al. (2012) and Lozano and Gutiérrez (2014), instead of a production to marketing/sales process, we adopt a *provision−delivery* model. For purposes of our current chapter, we will only measure each activity within the *provision−delivery* separately mainly to demonstrate the standard DEA model. In Chapter 7, we will then merge these two activities into one termed as *provision−delivery* model under a network DEA model.

3.2.1 Provision model

In airline operations, its purpose is to generate seats and cargo space for the purpose of conveying passengers and freight, respectively. From the extant airline literature, the outputs for airline service provision comprise ASK and ATK.[1]

According to the definitions of the International Civil Aviation Organization (ICAO) drawn from https://www.icao.int/Search/pages/results.aspx?k=glossary:

> *ASK is the "sum of the products obtained by multiplying the number of passenger seats available for sale on each flight stage by the stage distance. It excludes seats not available for the carriage of passengers because of the extra mass of fuel required or other payload restrictions".*

> *ATK is the "sum of the products obtained by multiplying the number of tonnes available for the carriage of revenue load (passengers, freight and mail) on each flight stage by the stage distance".*

In our *provision* model, we consider only one output, ATK, because the definition encapsulates both passengers and freight, whereas ASK only accounts for passengers. Depending on the focus of the activity within the supply chain model, one may consider other types of output indicators, such as RPK and revenue tonne kilometres (RTK). These output indicators, however, do not fall under the production framework described before because they are heavily dependent on demand-side conditions. In fact, these output indicators reflect the revenue side of the supply chain airline model and fall under the sales or marketing production side (Lu et al., 2012; Lozano and Gutiérrez, 2014).

To produce ATK, we consider two inputs—the number of maintenance employees and value of assets (property, plant and equipment) because they go directly into the production framework. The value of assets is used as it captures the values of equipment and other resources as well as the values of aircraft, which vary in aircraft size. The value of expenses on operating costs is

---

1. Includes both domestic and international.

also an input of the *provision* model. According to the definition of ICAO, carrier operating expenses are subdivided into direct and indirect operating costs:

> *Direct operating costs are costs incurred in operating the aircraft which cover the following main accounts: flight operations, flight equipment maintenance and overhaul, flight equipment depreciation, and user charges.*

The aforementioned activity reflects the *provision* model, unfortunately the World Air Transport Statistics eDataSeries does not provide data in terms of direct and indirect expenses, and thus, we do not use this variable in our *provision* model. Data for our inputs and output used in the *provision* model are presented in Appendix A Table A.2.

Besides determining the inputs and output in the *production* model, there is also the issue of determining the number of inputs, outputs, and decision-making units (DMUs). As noted in Sarkis (2007), the choice and number of inputs and outputs, and DMUs determine the quality of discrimination that can exist between efficient and inefficient firms. Table 3.1 presents a select list of studies and their rule of thumb regarding the minimum number of DMUs and x's and y's required for analysis to provide sufficient discrimination.

Cook et al. (2014, p. 2), however, argue that 'such a rule is neither imperative, nor does it have a statistical basis, but rather is often imposed for convenience'. Nonetheless, they feel that it is still important to consider all relevant inputs and outputs for DEA studies. For our *provision* model, the use of 2 inputs, 1 output, and 16 DMUs satisfies the minimum condition in the studies listed in Table 3.1 and that the productivity model would provide sufficient discrimination.

TABLE 3.1 Select studies and the rule of thumb ($n = 16$).

		Min. No. of DMUs required	
	Rule of thumb	Provision ($x = 2$, $y = 1$)	Delivery ($x = 3$, $y = 2$)
Boussofiane et al. (1991)	$X \times Y$	2	6
Golany and Roll (1989)	$2(X + Y)$	6	10
Banker et al. (1989); Bowlin (1998)	$3(X + Y)$	9	15
Friedman and Sinuany-Stern (1998)	$(X + Y) < n/3$	$3 < 4.33$	$5 < 5.33$
Dyson et al. (2001)	$2(X \times Y)$	4	12

3.2.2 Delivery model

For the delivery model, we use three inputs ATK, fuel burn and flight personnel, which are used to convey passengers and freight over distance. Data, except for fuel burn, are drawn from WATS (2017). Fuel burn is drawn from RDC Aviation. Flight personnel comprise pilots, copilots, other cockpit personnel and cabin crew. The result from consumption of these inputs is the output 'TKP'. According to the definition of ICAO, TKP is

the sum of the product obtained by multiplying the number of total tonnes of revenue load (passengers, freight and mail) carried on each flight stage by the stage distance.

TKP in ICAO reports passenger TKP and freight and mail TKP separately.

As noted in Coelli et al. (1999) that TKP best reflects the ticketing and marketing aspects of airline business rather than their actual flying operations. We agree and disagree to their statement because while kilometres performed is determined by demand side linked with sales and marketing, it is also part of airline operations and determined by fuel burn and number of flight personnel. Nonetheless, such demand variable should be factored into the production model as a nondiscretionary variable and considered a source of (in)efficiency, which we will examine in Chapter 8.

Our delivery model thus considers 3 inputs and 2 outputs and satisfies the minimum requirements of number of DMUs as specified in Table 3.1. Input and output data for the *delivery* model are provided in the first five columns of Appendix A Table A.3. Data for columns 6−10 are the unit prices with respect to the inputs and outputs mentioned before and are described in more detail in the following section. There are other measurements of inputs worth mentioning that need clarification before deciding on whether they are appropriate for our model. A number of planes as an input have been used in several studies, such as Good et al. (1995), Barbot et al. (2008), Barros and Peypoch (2009) and Nolan et al. (2014). We do not use this variable because the number of planes does not discriminate between large and small planes and provide no information of the capacity of a plane nor the distance travelled, which are essential inputs. Kilometres flown and hours flown are possible inputs. As we have already considered TKP, which already incorporates the distance flown factor, kilometres flown as an input becomes redundant. Hours flown is irrelevant in our model because our focus is not time crucial. Hours flown would be a useful input under circumstances where the measurement relates to quality of service (i.e. flight delays and cabin crew fatigue).

3.2.3 Cost and revenue efficiency model

We expand the analysis of the *delivery model* to include the unit price data (columns 6−10) as shown in Appendix A Table A.3 to illustrate the cost and revenue efficiency models. The unit prices of Appendix A Table A.3 are in US$ and deflated using purchasing power parities (PPPs) of the World Bank's World Development Indicators obtained from https://databank.worldbank.org/source/world-development-indicators#. Unit prices may either be obtained directly or indirectly. In our case, unit price of fuel (w2) was drawn from each airline's annual reports and were converted to a common unit (i.e. US$ per tonne). The following unit prices were derived by taking ratios using data from WATS (2017). Unit price of ATK (w2) is the ratio of assets (property, plant and equipment) to ATK. Unit price of airfare (p1) is the ratio of operating revenue passenger to passenger TKP Unit price of freight and mail (p2) is the ratio of operating revenue cargo to freight and mail TKP. Finally, the unit price of flight personnel (w1) is the ratio of wages, salaries and benefits to the total number of flight personnel. This ratio was derived using data from airline annual reports and WATS (2017). Wages, salaries and benefits drawn from annual reports were aggregated and not disaggregated into flight personnel. So we took the proportion of flight personnel to total employees (from WATS, 2017) and multiplied this to the total wages, salaries and benefits (from annual reports) and then deflated to US$ via PPP to derive a unit price proxy for flight personnel.

3.3 DEA models

DEA, based on linear mathematical programming, constructs a nonparametric piecewise frontier from a given data set. It essentially measures efficiency of DMUs relative to the frontier. Frontier efficiency concept, first elucidated in Farrell's (1957) seminal work, did not receive widespread attention until the work of Charnes et al. (1978). Charnes, Cooper and Rhodes (1978), henceforth CCR, were the first to introduce the term DEA and proposed an input-oriented CRS model.

3.3.1 CCR model

Suppose there are n DMUs, each using m inputs to produce u outputs. The envelopment form of the input-oriented constant returns to scale (CRS) model is expressed as

$$\min_{\theta,\lambda} \theta$$
$$\text{subject to} \quad \theta x_0 - X\lambda \geq 0, -y_0 + Y\lambda \geq 0, \quad (3.1)$$
$$\lambda \geq 0,$$

where x_0 is a vector of inputs, y_0 is the vector of outputs, X is the input matrix $(n \times m)$ and Y is the output matrix $(n \times u)$ for all DMUs. λ is a vector of intensity variables denoting linear combinations of DMUs, and θ is a scalar of efficiency scores for DMU$_0$'s. A measure of $\theta_0 = 1$ indicates that the DMU is technically efficient, whereas it is inefficient if $\theta_0 < 1$. This linear programming problem must be solved n times, once for each DMU in the sample.

3.3.2 BCC model

The CCR model was further developed by Banker et al. (1984), (henceforth BCC) by introducing a variable returns to scale (VRS) DEA model. The CRS model is then modified into a VRS model by imposing the convexity constraint, $\Sigma \lambda_0 = 1$ (i.e. sum of the intensity variables, λ's, equal to one). The BCC model is expressed as

$$\min_{\theta, \lambda} \theta$$
$$\text{subject to} \quad \theta x_0 - X\lambda \geq 0, -y_0 + Y\lambda \geq 0, \tag{3.2}$$
$$\lambda \geq 0,$$
$$\Sigma \lambda_0 = 1$$

The notable difference between CRS and VRS is that under CRS, a DMU may be benchmarked against other DMUs that are substantially larger (smaller) that it. Under VRS, the convexity constraint ensures that inefficient firms are benchmarked against firms of similar operational size.

3.3.3 Cost minimization model

If input prices are available, one can derive the allocative efficiency score via cost minimization input-oriented model as defined in Eq. (3.3).

$$\min_{\lambda, x_0^*} w_0' x_0^*,$$
$$\text{subject to} \quad x_0^* - X\lambda \geq 0, -y_0 + Y\lambda \geq 0, \tag{3.3}$$
$$\lambda \geq 0,$$

where w_0' is a vector of input prices for the 0-th firm and x_0^* (calculated via linear programming) is the cost minimizing vector of input quantities for the 0-th DMU. Cost efficiency (CE) or economic efficiency[2] of the 0-th DMU is calculated as

$$CE = \frac{w_0' x_0^*}{w_0' x_0} \tag{3.4}$$

which is the ratio of minimum cost to the observed cost for 0-th DMU. Allocative efficiency is then derived as a residual of CE over TE:

$$AE = \frac{CE}{TE} \tag{3.5}$$

---

2. See Coelli et al. (1998) for a discussion and illustration of Farrell's (1957) efficiency measurement, which showed that under CRS, economic efficiency = technical efficiency × allocative efficiency.

3.3.4 Revenue maximization model

If output prices are available, one can derive the allocative efficiency score via revenue maximization. Maintaining the use of input-oriented model, we define the revenue maximization model in Eq. (3.6) as

$$\max_{\lambda, y_0^*} p_0' y_0^*,$$
$$\text{subject to} \quad x_0 - X\lambda \geq 0, -y_0^* + Y\lambda \geq 0, \tag{3.6}$$
$$\lambda \geq 0,$$

where p_0' is a vector of output prices for the 0-th DMU and y_0^* (calculated via linear programming) is the revenue maximizing vector of output quantities for the 0-th DMU. Revenue efficiency (RE) or economic efficiency of the 0-th DMU is calculated as

$$\text{RE} = \frac{p_0' y_0}{p_0' y_0^*} \tag{3.7}$$

which is the ratio of observed revenue to the maximum revenue for 0-th DMU. Allocative efficiency is then derived as a residual of RE over TE:

$$\text{AE} = \frac{\text{RE}}{\text{TE}} \tag{3.8}$$

3.4 R package

The software used in this book is R. We provide the scripts for each DEA model appropriate for airline services. The main aim is to provide the scripts to assist readers who have limited knowledge in coding the models but mainly wish to apply DEA, and not reinvent the wheel.

There are a number of packages available on R that run DEA. These include 'Benchmarking' developed by Bogetoft and Otto (2011, 2015), 'deaR' developed by Coll-Serrano et al. (2020), 'productivity' developed by Dakpo et al. (2018), 'DJL' developed by Lim (2021), 'TFDEA' developed by Shott and Lim (2015), 'rDEA' developed by Simm and Besstremyannaya (2016), and 'FEAR' developed by Wilson (2008). We will occasionally employ one or more of the aforementioned packages in this chapter and also include other R packages in the coming chapters and other R scripts that are not available from any of the R packages.

3.5 R script for DEA, results and interpretation of results

Prior to preparing the DEA script, the data (as shown in Appendix A Table A.2) must be saved in text format. The first row shows the information such as DMU and type of variable (e.g. $x1$, $y1$). The first column reads the DMUs, and subsequent columns read the input(s) and output(s). We name our

TABLE 3.2 RStudio 'files/plots/packages/help' window.

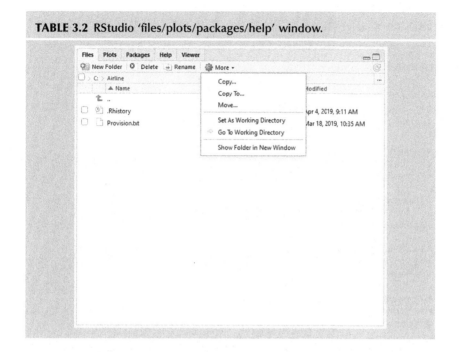

file 'Provision.txt' and save it in folder C:/Airline. As such, R script for DEA has to read the data as follows:

read.table(file="C:/Airline/Provision.txt",header=T,sep="\t")

In addition, one may be required to set the working directory found under the 'Files/plots/packages/help' window. Locate the relevant directory and then select 'Set as working directory' as shown in Table 3.2.

Alternatively, one could type in the command

setwd("C:/Airline")

in the 'Console' window to set up the working directory.

3.5.1 R script for DEA (Charnes et al. 1978) CCR model

The R script for DEA is shown in Table 3.3 and is based on the 'Benchmarking' package developed by Bogetoft and Otto (2011, 2015).

The aforementioned script is written for airlines under CRS and input orientation and is the CCR model developed by Charnes et al. (1978). We assume CRS because our list of airlines are mainstream airlines and assumed to be operating at roughly the same operational scale. One could test the underlying technology on whether it exhibits CRS or VRS as discussed in

TABLE 3.3 R script for DEA ('provision' model).

```
## load Benchmarking
library(Benchmarking)

## load data
data <- read.table(file="C:/Airline/Provision.txt", header=T,sep="\t")
attach(data)

## y refers to outputs, x refers to inputs
y <- cbind(y1)
x <- cbind(x1,x2)

# To derive Technical efficiency
te <- dea(x,y,RTS="crs",ORIENTATION="in")
data.frame(dmu,te$eff)

# To obtain lambdas (weights) of peers for inefficient firms
weights<- lambda(te)
data.frame(dmu,weights)

# To calculate slacks
sl <- slack(x,y,te)
data.frame(dmu,sl$slack,sl$sx,sl$sy)

# To calculate proportionate or radial movements
radial<-excess(te,x,y)
data.frame(dmu,radial)
```

Section 3.6.3. But at this stage, we assume CRS. If one wishes to change to VRS, then RTS should equal to 'vrs' in Table 3.3. We consider an input orientation because of the assumption that airlines are unable to change their output (i.e. the number of seats and cargo space) and thus can only adjust their inputs. Furthermore, outputs in the *delivery* model are mainly driven by demand side, which is beyond the control of airlines. To change to output orientation, one simply needs to change ORIENTATION="in" to ORIENTATION="out" in Table 3.3.

3.5.2 Interpretation of DEA (CCR) results for the 'provision' model

Table 3.4 presents the DEA efficiency scores for the input-oriented under CRS. Airlines having an efficiency score of less than 1.00 are considered inefficient. The results show that Singapore Airlines and United Airlines operated efficiently in 2009. To interpret the efficiency score, using KLM as an example, its efficiency score of 0.944 (rounded upwards) means that it is operating at 94.4% efficiency and that it can still improve efficiency by reducing consumption of its inputs by 5.6% (i.e. 100%−94.4% = 5.6%). As DEA

TABLE 3.4 Output report of DEA scores (CRS) based on Table 3.3 script.

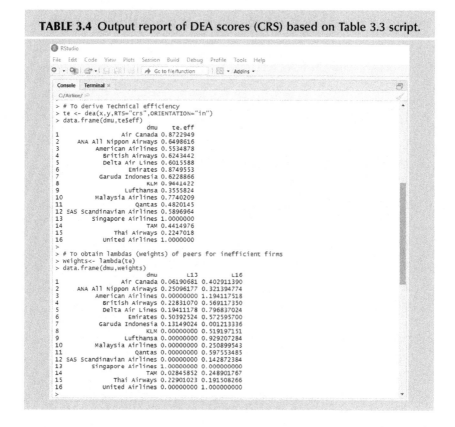

efficiency scores are relative scores derived from peer assessment with other airlines, it is possible to determine how inefficient airlines can improve by comparing themselves with efficient airlines (i.e. benchmark airlines).

Benchmark weights are also generated from the R script of Table 3.3, and the results therefrom are presented in Table 3.4 and referred to as lambdas (λ) in the generated output. In our case, L13 and L16 are lambdas (λ) for Singapore Airlines and United Airlines, respectively.

The lambdas (λ) are weights relative to the benchmark airlines to determine the efficiency improvements needed for the inefficient airlines. For example, KLM's benchmark (percentage of target needed) against Singapore Airlines is 0% and against United Airlines is 51.9%. This indicates that KLM's operation is more similar to United Airlines and should adjust its input consumption using United Airlines as its benchmark. We observe that all inefficient airlines are benchmarked against United Airlines, whereas only 8 inefficient airlines are benchmarked against Singapore Airlines suggesting some differences in operations between the efficient airlines. It is worth noting that the benchmark

TABLE 3.5 Output report of DEA scores (VRS) based on Table 3.3 script.

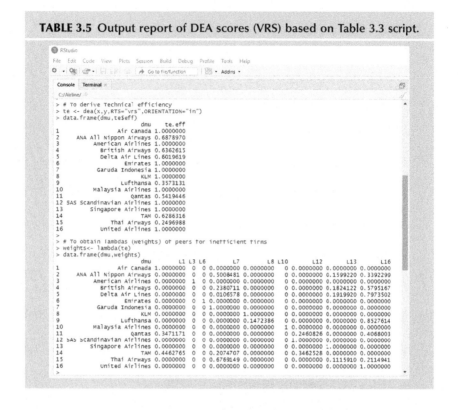

for all inefficient airlines, the weights (i.e. lambdas (λ)), may not necessarily equal to 1.00 (i.e. 100%), for example, Emirates. This is because the convexity constraint, $\Sigma \lambda_0 = 1$ (i.e. sum of the intensity variables, λ's, equal to one), is not imposed in the CRS model.

It is important to note here that efficiency scores based on CRS include scale efficiency. If we calculate the efficiency scores under VRS as shown in Table 3.5, we only measure the pure technical efficiency of airlines. Under VRS, we observe that there are now 9 efficient airlines which suggest the significance scale efficiency can have on the overall efficient levels. If we look at KLM, we observe that it is now operating efficiently suggesting how VRS can provide a different outcome to CRS. If we use Lufthansa to illustrate between CRS and VRS, under CRS it was only benchmarked against United Airlines. But under VRS, it is now benchmarked against KLM (14.7%) and United Airlines (85.3%). Note $14.7 + 85.3 = 100\%$. While this might suggest that Lufthansa can improve its performance by benchmarking with United Airlines and to some extent KLM, because the measure is under VRS, scale efficiency needs to be considered to determine economic efficiency. This is because CRS technical efficiency measure is decomposed into 'pure' technical

TABLE 3.6 Radial movements and slack movements (CRS) for the 'provision' model.

```
RStudio
File   Edit   Code   View   Plots   Session   Build   Debug   Profile   Tools   Help
      Go to file/function           Addins

Console   Terminal
C:/Airline/
> # To calculate slacks
> sl <- slack(x,y,te)
> data.frame(dmu,sl$slack,sl$sx,sl$sy)
                        dmu sl.slack        sx1 sx2 sy1
1                 Air Canada    FALSE   0.000000   0   0
2         ANA All Nippon Airways FALSE   0.000000   0   0
3           American Airlines    TRUE 308.850939   0   0
4             British Airways    FALSE   0.000000   0   0
5             Delta Air Lines    FALSE   0.000000   0   0
6                    Emirates    FALSE   0.000000   0   0
7            Garuda Indonesia    FALSE   0.000000   0   0
8                         KLM    TRUE 2036.581223   0   0
9                  Lufthansa    TRUE 2488.780219   0   0
10          Malaysia Airlines    TRUE 1683.211401   0   0
11                    Qantas    TRUE   1.536902   0   0
12  SAS Scandinavian Airlines    TRUE 507.462429   0   0
13          Singapore Airlines   FALSE   0.000000   0   0
14                        TAM    FALSE   0.000000   0   0
15               Thai Airways    FALSE   0.000000   0   0
16             United Airlines   FALSE   0.000000   0   0
>
> # To calculate proportionate or radial movements
> radial<-excess(te,x,y)
> data.frame(dmu,radial)
                        dmu          x1           x2
1                 Air Canada   292.82774 9.216630e+05
2         ANA All Nippon Airways 907.20854 5.130167e+06
3           American Airlines  4966.55557 1.174774e+07
4             British Airways  1737.03225 7.242425e+06
5             Delta Air Lines  2640.86814 9.306391e+06
6                    Emirates   432.27960 2.605635e+06
7            Garuda Indonesia    38.46557 1.786057e+06
8                         KLM   270.91033 3.745937e+05
9                  Lufthansa 12756.89171 2.053627e+07
10          Malaysia Airlines   850.13354 8.933001e+05
11                    Qantas  3146.24368 7.830978e+06
12  SAS Scandinavian Airlines   839.89150 1.212291e+06
13          Singapore Airlines    0.00000 1.982692e-08
14                        TAM  1557.66322 4.643426e+06
15               Thai Airways  3581.87754 2.569700e+07
16             United Airlines    0.00000 0.000000e+00
>
```

efficiency and scale efficiency. We will examine the scale efficiency in more detail in this section under '*scale efficiency*'.

3.5.2.1 Interpreting radial (proportionate) and slack movements

Table 3.6 details the potential improvements for each inefficient airline needed to achieve overall efficiency under CRS. It shows the proportionate movement (or radial movement) and slack movement (if available) required for inefficient airlines to change the inputs and/or outputs to become efficient. Note that the reported estimates for radial movement and slack movement in Table 3.6 are generally negative for inputs and positive for outputs. $sx1$ and $sx2$ are the slack movements for $x1$ and $x2$, respectively.[3]

---

3. Note that although not evident in our input-orientation model, it is still possible for both input and output slacks to coexist.

Using American Airlines as an example, we observe that it can improve its overall efficiency by reducing its number of maintenance employees by 4967−6156 and assets by $11,747,736,866−$14,562,263,134, and then a further reduction in the number of maintenance employees by 309 to 5848 (= 6156−309). The meaning of the radial and slack movements for American Airlines can be illustrated in Fig. 3.1.

Fig. 3.1 shows the frontier (FF′) under input orientation CRS model. It also depicts the technical efficiency of American Airlines and United Airlines (as the benchmark airline of American Airlines), radial and slack movements from Tables 3.4 and 3.6.

United Airlines is on the frontiers as it is efficient, whereas American Airlines is beyond the frontier indicating inefficiency. American Airlines can improve efficiency by first cutting back on the use of inputs $x1$ and $x2$ (as described earlier) so that it moves to the point American Airlines. This is the radial (or proportionate movement). But at American Airlines, it is on the frontier but could still improve on its efficiency by cutting back on $x1$ (i.e. number of maintenance employees) to reach the point where United Airlines is. This is known as the input slack.

3.5.2.2 Scale efficiency

At times, we wish to determine if the size of operations may be attributing to inefficiency. In the case of our airline example, this would be a meaningful exercise as one would expect airlines to be operating at different levels of

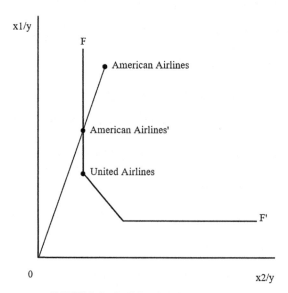

FIGURE 3.1 Radial and slack movements.

TABLE 3.7 R script for scale efficiency (provision model).

```
## Load library(deaR)
library(deaR)

## Load data
data <- read.table(file="C:/Airline/Provision.txt", header=T,sep="\t")
attach(data)
data_dea <- read_data(data,ni = 2,no = 1)

tecrs <- model_basic(data_dea, orientation = "io", rts = "crs")
tevrs <- model_basic(data_dea, orientation = "io", rts = "vrs")
tenirs <- model_basic(data_dea, orientation = "io", rts = "nirs")

tecrs <- efficiencies(tecrs)
tevrs <- efficiencies(tevrs)
tenirs <- efficiencies(tenirs)

## Input oriented scale efficiency
scale_eff = round((tecrs/tevrs),6)

data.frame(tecrs,tevrs,tenirs,scale_eff)
```

capacity. To determine if a firm is operating at the optimal scale, we measure its scale efficiency by taking the ratio of efficiency scores of CCR/BCC (Banker, 1984). R script for scale efficiency is shown in Table 3.7 using deaR package developed by Coll-Serrano et al. (2020), although one could easily calculate scale efficiency using excel spreadsheet. As pointed out by Golany and Roll (1989), CCR efficiency, under the assumption of CRS, is decomposed of pure technical efficiency and scale efficiency. BCC under VRS measures only pure technical efficiency and excludes any scale effects.

Table 3.8 presents the efficiency scores under CRS, VRS and NIRS (non-increasing returns to scale), as well as scale efficiency. We compute the NIRS to determine an airline's RTS. As Coelli et al. (2005) notes, if a firm's NIRS technical efficiency = VRS technical efficiency, then it is exhibiting decreasing returns to scale (DRS). If a firm's NIRS technical efficiency ≠ VRS technical efficiency, then it is exhibiting increasing returns to scale (IRS). If a firm's CRS technical efficiency = VRS technical efficiency, then it is exhibiting constant returns to scale (CRS). These descriptors are more easily discerned in Fig. 3.2.

In Fig. 3.2, the NIRS range is 0AC, whereas VRS is BAC, which indicates that firms which fall within the AC range, regardless of VRS or NIRS (because the technical efficiency scores are the same), would be exhibiting DRS. (Note as X increases, Y increases but at a diminishing rate). For the range BA (under

TABLE 3.8 CRS, VRS, NIRS technical efficiency and scale efficiency scores for provision model.

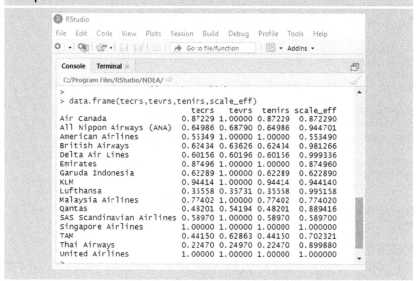

```
> 
> data.frame(tecrs,tevrs,tenirs,scale_eff)
                                tecrs    tevrs   tenirs  scale_eff
Air Canada                    0.87229 1.00000 0.87229  0.872290
All Nippon Airways (ANA)      0.64986 0.68790 0.64986  0.944701
American Airlines             0.55349 1.00000 1.00000  0.553490
British Airways               0.62434 0.63626 0.62434  0.981266
Delta Air Lines               0.60156 0.60196 0.60156  0.999336
Emirates                      0.87496 1.00000 1.00000  0.874960
Garuda Indonesia              0.62289 1.00000 0.62289  0.622890
KLM                           0.94414 1.00000 0.94414  0.944140
Lufthansa                     0.35558 0.35731 0.35558  0.995158
Malaysia Airlines             0.77402 1.00000 0.77402  0.774020
Qantas                        0.48201 0.54194 0.48201  0.889416
SAS Scandinavian Airlines     0.58970 1.00000 0.58970  0.589700
Singapore Airlines            1.00000 1.00000 1.00000  1.000000
TAM                           0.44150 0.62863 0.44150  0.702321
Thai Airways                  0.22470 0.24970 0.22470  0.899880
United Airlines               1.00000 1.00000 1.00000  1.000000
```

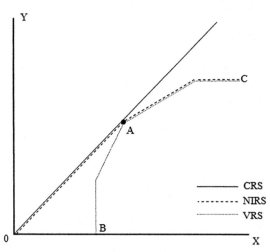

FIGURE 3.2   Returns to scale.

VRS) and 0A (under NIRS), the difference in their technical efficiency scores indicates firms in the BA range are exhibiting IRS. This is shown by the amount Y increases relative to X.

TABLE 3.9 R script for TE (CRS and VRS), scale efficiency, cost and revenue efficiency DEA (Delivery model).

```
## load Benchmarking
library(Benchmarking)

## load data
data <- read.table(file="C:/Airline/Delivery.txt",header=T,sep="\t")
attach(data)

## y refers to outputs, x refers to inputs, w refers to input prices, p refers to output prices
y <- cbind(y1,y2)
x <- cbind(x1,x2,x3)
w <- cbind(w1,w2,w3)
p <- cbind(p1,p2)

# To derive Technical efficiency
te_crs <- dea(x,y,RTS="crs",ORIENTATION="in")
te_vrs <- dea(x,y,RTS="vrs",ORIENTATION="in")

## Compute the Farrell input oriented scale efficiency
se = round((te_crs$eff /te_vrs$eff),5)

## Tabulate results
data.frame(dmu,crs=round(te_crs$eff,5),vrs=round(te_vrs$eff,5),se=round(se,5))

# To obtain lambdas (weights) of peers for inefficient firms
weights_te<- lambda(te_crs)
data.frame(dmu,weights_te)

# To calculate slacks
sl <- slack(x,y,te_crs)
data.frame(dmu,sl$slack,sl$sx,sl$sy)

# To calculate proportionate or radial movements
radial<-excess(te_crs,x,y)
data.frame(dmu,radial)

# Cost efficiency
xopt <- cost.opt(x,y,w, RTS="crs")
cobs <- x*w
copt <- xopt$x*w
ce <- rowSums(copt)/rowSums(cobs)

# To obtain lambdas (weights) of peers for inefficient firms
weights_ce<- lambda(xopt)
data.frame(dmu,weights_ce)

# Allocative efficiency for CE
aecost <- ce/te_crs$eff

# Revenue efficiency
yopt <- revenue.opt(x,y,p,RTS="crs")
robs <- y*p
```

Identifying the input prices (w1, w2 and w3) and output price (p1 and p2).

Performing cost efficiency under CRS. Copt is the numerator and cobs is the denominator of the cost efficiency model (3.4).

Performing revenue efficiency under CRS. Robs is the numerator and ropt is the denominator of the revenue efficiency model (3.7).

Continued

TABLE 3.9 R script for TE (CRS and VRS), scale efficiency, cost and revenue efficiency DEA (Delivery model).—cont'd

```
ropt <- yopt$y*p
re <- rowSums(robs)/ rowSums(ropt)

# Allocative efficiency for RE
aerev <- re/te_crs$eff

# To obtain lambdas (weights) of peers for inefficient firms
weights_re<- lambda(yopt)
data.frame(dmu,weights_re)

#Cost efficiency, revenue efficiency, technical efficiency, allocative efficiency (cost and revenue)
data.frame(dmu,ce,re,te_crs$eff,aecost,aerev)
```

3.5.3 R script for DEA ('delivery' model)

For the *delivery* model, we focus on the cost and revenue efficiency models but still include the technical efficiency models. The script displayed in Table 3.9 is similar to Table 3.3; only this time, we include scripts for measuring cost and revenue efficiency. We then use the data set shown in Appendix A Table 3.3 (named as 'Delivery') to generate the technical efficiency, cost efficiency, revenue efficiency and allocative efficiency scores of cost and revenue efficiency. Cost and revenue efficiency models require price data which we define as w for input prices and p for output price in the R script of Table 3.9.

3.5.4 Interpretation of DEA results for the 'delivery' model

In this section, we first describe and interpret the results for technical efficiency under CRS, VRS and scale efficiency and also discuss the radial and slack movements. We then discuss the cost and revenue efficiency scores, allocative efficiency for cost and revenue efficiency. Table 3.10 presents the DEA technical efficiency (TE) scores and benchmark Airlines (weights) based on the R script of Table 3.9.

In the *delivery* model, KLM, Qantas and Singapore Airlines are the only technically efficient airlines under CRS and VRS. Of these three airlines, KLM and Qantas are benchmarked with most of the airlines suggesting similar organizational structure. Some airlines are benchmarked against one or two airlines. For example, only Emirates is benchmarked against all three efficient airlines—KLM (105%), Qantas (13.5%) and Singapore Airlines (28.5%). Note that the $\Sigma \lambda_0$ does not equal to 1 $(0.011 + 0.584 = 0.595)$ because the convexity constraint, $\Sigma \lambda_0 = 1$ (i.e. sum of the intensity variables, λ's, equal to

TABLE 3.10 Technical efficiency scores (CRS, VRS), scale efficiency and benchmark weights.

```
R RStudio

File   Edit   Code   View   Plots   Session   Build   Debug   Profile   Tools   Help
  ▾   | ⚙ | ▾ | ⌂ | ⊞ | ⊟ |   ↗ Go to file/function      | ⊞ ▾ Addins ▾

Console   Terminal ×
C:/Program Files/RStudio/NDEA/ ⇌
> # To derive Technical efficiency
> te_crs <- cea(x,y,RTS="crs",ORIENTATION="in")
> te_vrs <- cea(x,y,RTS="vrs",ORIENTATION="in")
>
> ## Compute the Farrell input oriented scale efficiency
> se = round((te_crs$eff /te_vrs$eff),5)
>
> ## Tabulate results
> data.frame(dmu,crs=round(te_crs$eff,5),vrs=round(te_vrs$eff,5),se=round(se,5))
                         dmu      crs      vrs       se
1                  Air Canada  0.90979  0.91959  0.98934
2           ANA All Nippon Airways  0.72944  0.77130  0.94572
3             American Airlines  0.97818  1.00000  0.97818
4               British Airways  0.90965  0.95256  0.95495
5                Delta Air Lines  0.98870  1.00000  0.98870
6                       Emirates  0.97791  1.00000  0.97791
7               Garuda Indonesia  0.93331  1.00000  0.93331
8                            KLM  1.00000  1.00000  1.00000
9                     Lufthansa  0.94479  1.00000  0.94479
10             Malaysia Airlines  0.97650  1.00000  0.97650
11                        Qantas  1.00000  1.00000  1.00000
12       SAS Scandinavian Airlines  0.96911  1.00000  0.96911
13              Singapore Airlines  1.00000  1.00000  1.00000
14                            TAM  0.87668  0.88955  0.98553
15                  Thai Airways  0.88186  0.89623  0.98396
16                United Airlines  0.93561  0.96873  0.96582
>
> # To obtain lambdas (weights) of peers for inefficient firms
> weights_te<- lambda(te_crs)
> data.frame(dmu,weights_te)
                         dmu         L8         L11         L13
1                  Air Canada  0.2834347  0.4361980  0.000000000
2           ANA All Nippon Airways  0.5833886  0.0000000  0.000000000
3             American Airlines  0.4487664  1.4649020  0.000000000
4               British Airways  0.8255635  0.4035888  0.000000000
5                Delta Air Lines  0.6805262  0.9623535  0.000000000
6                       Emirates  1.0515964  0.1350288  0.285418836
7               Garuda Indonesia  0.0000000  0.1523000  0.000000000
8                            KLM  1.0000000  0.0000000  0.000000000
9                     Lufthansa  1.6777490  0.0000000  0.009313509
10             Malaysia Airlines  0.2198326  0.0000000  0.178695292
11                        Qantas  0.0000000  1.0000000  0.000000000
12       SAS Scandinavian Airlines  0.0000000  0.2317087  0.000000000
13              Singapore Airlines  0.0000000  0.0000000  1.000000000
14                            TAM  0.0000000  0.3957448  0.000000000
15                  Thai Airways  0.4225271  0.1630119  0.000000000
16                United Airlines  0.7158512  0.9437559  0.000000000
>
```

one), is not imposed in the CRS model; hence the weights (i.e. lambdas (λ)) may not necessarily equal to 1.00 (i.e. 100%). The λ's for Emirates suggest that for the airline to improve its performance, the most important target is KLM as its (efficient) combination of inputs and outputs is closest to it.

To determine an airline's operational size, we take the ratio of efficiency scores of CCR/BCC to derive scale efficiency as shown in Table 3.11 based on the R script of Table 3.7; only this time, changes to the number of inputs and outputs have to be made for the line:

data_dea < - read_data(data, ni = 3, no = 2)

TABLE 3.11 CRS, VRS, NIRS technical efficiency and scale efficiency scores for delivery model.

```
R  RStudio

File  Edit  Code  View  Plots  Session  Build  Debug  Profile  Tools  Help

  ▾  ▾    ▾             ⟶ Go to file/function       ▾  Addins ▾

Console  Terminal ×                                                          ⎙

C:/Program Files/RStudio/NDEA/

>
> data.frame(tecrs,tevrs,tenirs,scale_eff)
                             tecrs    tevrs   tenirs  scale_eff
Air Canada                 0.90979  0.91959  0.90979   0.989343
ANA All Nippon Airways     0.72944  0.77130  0.72944   0.945728
American Airlines          0.92818  1.00000  1.00000   0.928180
British Airways            0.90965  0.95256  0.95256   0.954953
Delta Air Lines            0.98870  1.00000  1.00000   0.988700
Emirates                   0.97791  1.00000  1.00000   0.977910
Garuda Indonesia           0.93331  1.00000  0.93331   0.933310
KLM                        1.00000  1.00000  1.00000   1.000000
Lufthansa                  0.94479  1.00000  1.00000   0.944790
Malaysia Airlines          0.97650  1.00000  0.97650   0.976500
Qantas                     1.00000  1.00000  1.00000   1.000000
SAS Scandinavian Airlines  0.96911  1.00000  0.96911   0.969110
Singapore Airlines         1.00000  1.00000  1.00000   1.000000
TAM                        0.87668  0.88955  0.87668   0.985532
Thai Airways               0.88186  0.89623  0.88186   0.983966
United Airlines            0.93561  0.96873  0.96873   0.965811
>
```

We observe that for British Airways, it is benchmarked against KLM and Qantas (from Table 3.10). As its VRS=NIRS (see Table 3.11), it exhibits DRS, which suggests that its operations are too large and it has to cut back on its inputs.

To determine the amount of inputs British Airways has to cut back, we look at the radial and slack movements presented in Table 3.12.

From Table 3.12, British Airways will need to cut back on flight personnel by the sum of radial movement and slack movement equaling 4969.08 (i.e. 1496.4124 + 3472.672 = 4969.08), cut back on ATK by radial movement of 1,933,562,135 and cut back on fuel burn (i.e. consumption) by the sum of radial movement and slack movement equaling 1,531,007.91 (i.e. 479,236.08 + 1,051,771.83 = 1,531,007.91). Some airlines such as Air Canada not only have to cut back on some inputs but also increase outputs (i.e. y2).

3.5.5 Cost and revenue efficiency model

In this section, we present the cost and revenue efficiency scores based on the *delivery model*. Data for cost and revenue efficiency measurement are presented in Appendix A Table A.3. Using the R script in Table 3.9, we generate the cost efficiency (CE), revenue efficiency (RE) scores and allocative efficiency (AE) scores for CE and RE and present them in Table 3.13. Note that RE scores will be the same as TE scores if only one output was considered meaning that AE for revenue maximization will equal to 1.00. But because we

TABLE 3.12 Radial movements and slack movements (CRS) for delivery model.

```
> # To calculate slacks
> sl <- slack(x,y,te_crs)
> data.frame(dmu,sl$slack,sl$sx,sl$sy)
                       dmu sl.slack        sx1        sx2       sx3 sy1         sy2
1               Air Canada     TRUE     0.0000         0  549791.66   0  1147496047
2       ANA All Nippon Airways   TRUE     0.0000 1906947516   98424.57   0   328792368
3         American Airlines     TRUE     0.0000         0 2051204.84   0  3262219447
4           British Airways     TRUE  3472.6720         0 1051771.83   0           0
5           Delta Air Lines     TRUE     0.0000         0 2169153.89   0  3639320873
6                  Emirates     TRUE     0.0000 2474786517       0.00   0           0
7          Garuda Indonesia     TRUE   189.7989         0  150576.93   0   117423532
8                       KLM    FALSE     0.0000         0       0.00   0           0
9                 Lufthansa     TRUE 17770.4634         0  329128.67   0           0
10        Malaysia Airlines     TRUE  1635.5033         0  275653.75   0           0
11                   Qantas    FALSE     0.0000         0       0.00   0           0
12 SAS Scandinavian Airlines    TRUE  1104.3480         0  342656.43   0   262883892
13        Singapore Airlines   FALSE     0.0000         0       0.00   0           0
14                       TAM    TRUE  1159.5200         0   51604.31   0   882422248
15             Thai Airways     TRUE  1098.3382         0  338390.64   0           0
16           United Airlines     TRUE     0.0000         0 2009400.20   0  3065706251
>
> # To calculate proportionate or radial movements
> radial<-excess(te_crs,x,y)
> data.frame(dmu,radial)
                       dmu        X1         X2         X3
1               Air Canada  753.4725 1.175371e+09  276126.24
2       ANA All Nippon Airways 1752.9690 3.972801e+09  691694.62
3         American Airlines 1659.1945 2.492722e+09  621597.72
4           British Airways 1496.4124 1.933562e+09  479236.08
5           Delta Air Lines  196.6884 3.083699e+08   83045.10
6                  Emirates  290.5468 6.045850e+08  104203.48
7          Garuda Indonesia  145.8422 1.890002e+08   45102.81
8                       KLM    0.0000 5.026236e-06       0.00
9                 Lufthansa 1837.9211 1.491183e+09  318013.44
10        Malaysia Airlines  122.9243 1.713689e+08   37760.94
11                   Qantas    0.0000 -6.170440e-05       0.00
12 SAS Scandinavian Airlines  125.0007 1.282962e+08   34237.03
13        Singapore Airlines    0.0000 1.181617e-05       0.00
14                       TAM  859.8067 9.668565e+08  248500.95
15             Thai Airways  871.1970 1.233551e+09  185656.25
16           United Airlines 1188.5937 1.871461e+09  492425.16
>
```

TABLE 3.13 CE, RE, TE and AE (cost and revenue) (CRS).

```
> #Cost efficiency, revenue efficiency, technical efficiency, allocative efficiency (cost and revenue)
> data.frame(dmu,ce,re,te crs$eff,aecost,aerev)
                       dmu        ce        re te_crs.eff    aecost     aerev
1               Air Canada 0.8266892 0.8747223  0.9097854 0.9088839 0.3614601
2       ANA All Nippon Airways 0.5895326 0.7186639  0.7294383 0.8082007 0.9852291
3         American Airlines 0.8682885 0.8840301  0.9281796 0.9354747 0.9524343
4           British Airways 0.8771554 0.8472696  0.9096533 0.9642744 0.3314203
5           Delta Air Lines 0.8911710 0.9242138  0.9887013 0.9013552 0.9347756
6                  Emirates 0.8836718 0.9732496  0.9779102 0.9036328 0.9952341
7          Garuda Indonesia 0.9143861 0.9099468  0.9333141 0.9797196 0.3749631
8                       KLM 1.0000000 1.0000000  1.0000000 1.0000000 1.0000000
9                 Lufthansa 0.9152506 0.8580747  0.9447873 0.9687372 0.3082200
10        Malaysia Airlines 0.9322642 0.7601395  0.9765008 0.9546989 0.7784320
11                   Qantas 1.0000000 1.0000000  1.0000000 1.0000000 1.0000000
12 SAS Scandinavian Airlines 0.8718328 0.9445422  0.9691051 0.8996267 0.9746540
13        Singapore Airlines 1.0000000 0.9424391  1.0000000 1.0000000 0.9424391
14                       TAM 0.8297074 0.4393470  0.8766804 0.9464195 0.5011484
15             Thai Airways 0.8582483 0.8329566  0.8818556 0.9732300 0.9445499
16           United Airlines 0.8394496 0.8865531  0.9356125 0.8972193 0.9466025
>
> |
```

TABLE 3.14 R script for shadow prices for DEA—CRS (delivery model).

```
## load Benchmarking
library(Benchmarking)

## load data
data <- read.table(file="C:/Airline/Delivery.txt",header=T,sep="\t")
attach(data)

## y refers to outputs, x refers to inputs
y <- cbind(y1,y2)
x <- cbind(x1,x2,x3)

# To derive the shadow prices (ux = shadow price of inputs; vy = shadow price of outputs)
sp <- dea(x,y,RTS="crs",ORIENTATION="in", DUAL=TRUE)
data.frame(sp$ux,sp$vy)
```

used two outputs, the RE and TE scores are not the same. We can derive AE for cost minimization residually by taking the ratio of CE to TE. We can also do this for revenue maximization by taking the ratio of RE to TE.

Here, we pause for a moment to comment on shadow prices. Shadow prices, derived using the R script of Table 3.14, uses DEA to solve the dual problem as outlined in Bogetoft and Otto (2011). In essence, DEA derives shadow prices exclusively from the quantity data. However, using DEA to solve the dual problem poses the problem of prices equalling to zero, leading to implausible findings and lacking economic meaning (Førsund, 2015). Kuosmanen et al. (2006) noted that besides shadow prices approaching zero, there are extreme cases where the shadow price approaches infinity and 'shadow prices obtained as the optimal solution of a DEA model tend to exhibit much variation across firms' (p. 736).

Nonetheless, we present the results on cost and revenue efficiency for the avid reader, using the first five columns of Appendix A Table A.3 and shadow prices derived therefrom, and present the results in Table 3.15.

The results between Tables 3.13 and 3.15 support Forsund (2015) that DEA based on shadow prices generates higher efficiency scores and implausible results (i.e. NaN: not a number). The table also supports Bogetoft and Otto (2011, p. 135) statement that 'the dualization thus supports the popular view that DEA puts everyone in the best possible light'.

3.6 Reliability of results

3.6.1 Bootstrapping DEA

A drawback of DEA is that it is a nonparametric estimator and does not accommodate for noise or random error. Hence, because it lacks statistical

TABLE 3.15 CE, RE, TE and AE (cost and revenue) (CRS) using shadow prices.

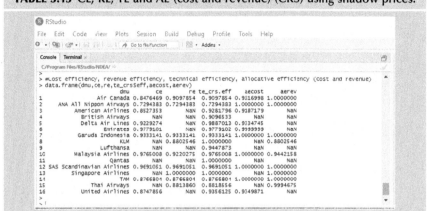

inference and is vulnerable to sampling variability, it raises the issue of reliability of the efficiency scores. In this section, we employ the bootstrap method of Simar and Wilson (1998) to derive bias-corrected estimates and statistical inference in the form of confidence intervals. We use Simar and Wilson's (1998) bootstrap method because it clearly defines a model of the

TABLE 3.16 R script for bias-corrected efficiency scores and confidence intervals (Simar and Wilson, 1998).

```
library(rDEA)

## load data
data <- read.table(file="C:/Airline/Provision.txt", header=T,sep="\t")
attach(data)

## y refers to outputs, x refers to inputs
Y <- cbind(y1)
X <- cbind(x1,x2)

firms=1:16

##Technical efficiency score
te_naive = dea(XREF=X, YREF=Y, X=X[firms,], Y=Y[firms,], model="input",
RTS="constant")

## Bias-corrected DEA scores and confidence intervals for bias-corrected DEA scores

te_robust = dea.robust(X=X[firms,], Y=Y[firms,], model="input",
RTS="constant", B=2000, alpha=0.05, bw_mult=0.014)

data.frame(dmu,te_naive$thetaOpt,te_robust$theta_hat_hat,te_robust
$bias,te_robust$theta_ci_low,te_robust$theta_ci_high)
```

> bw (bandwidth) is a smoothing parameter in sampling with reflection. "cv" or "bw.ucv" for cross-validation bandwidth, "silverman" or "bw.nrd0" for Silverman's (1986) rule, or h=0.014 as proposed in Simar and Wilson (1998).

TABLE 3.17 Efficiency scores, bias-corrected efficiency scores and confidence intervals (bandwidth $h = 0.014$).

```
RStudio
File  Edit  Code  View  Plots  Session  Build  Debug  Profile  Tools  Help
        Go to file/function          Addins

Console  Terminal
C:/Airline/
> ##Technical efficiency score
> te_naive = dea(XREF=X, YREF=Y, X=X[firms,], Y=Y[firms,], model="input", RTS="constant")
>
> ## Bias-corrected DEA scores and confidence intervals for bias-corrected DEA scores
>
> te_robust = dea.robust(X=X[firms,], Y=Y[firms,], model="input", RTS="constant", B=2000, alpha=0.05,
bw_mult=0.014)
>
> data.frame(dmu,te_naive$thetaopt,te_robust$theta_hat_hat,te_robust$bias,te_robust$theta_ci_low,te_ro
bust$theta_ci_high)
                               dmu  te_naive.thetaopt  te_robust.theta_hat_hat  te_robust.bias
1                        Air Canada          0.8722949                0.8064302      0.06586477
2              ANA All Nippon Airways          0.6498616                0.6068406      0.04302106
3                  American Airlines          0.5534878                0.4893530      0.06413472
4                   British Airways          0.6243442                0.5860207      0.03832351
5                   Delta Air Lines          0.6015588                0.5616640      0.03989481
6                          Emirates          0.8749553                0.8140592      0.06089610
7                  Garuda Indonesia          0.6228866                0.4005123      0.22237432
8                               KLM          0.9441422                0.8837667      0.06037555
9                        Lufthansa          0.3555824                0.3292108      0.02637152
10                 Malaysia Airlines          0.7740209                0.7283916      0.04562931
11                           Qantas          0.4820145                0.4212663      0.06074823
12          SAS Scandinavian Airlines          0.5896964                0.5507858      0.03891059
13                 Singapore Airlines          1.0000000                0.5685476      0.43145244
14                               TAM          0.4414976                0.4046633      0.03683425
15                      Thai Airways          0.2247018                0.2053494      0.01935240
16                   United Airlines          1.0000000                0.8737981      0.12620191
   te_robust.theta_ci_low  te_robust.theta_ci_high
1               0.7418092                 1.0212972
2               0.5647324                 0.7516808
3               0.4256062                 0.6533072
4               0.5486937                 0.7143038
5               0.5225535                 0.7035767
6               0.7545738                 1.0142679
7               0.1792276                 0.9624367
8               0.8238061                 1.1105558
9               0.3030529                 0.4175135
10              0.6830643                 0.9163352
11              0.3611397                 0.5690258
12              0.5121600                 0.6923677
13              0.1389953                 1.5983672
14              0.3686556                 0.5167007
15              0.1864125                 0.2616096
16              0.7490485                 1.1804119
>
> |
```

data-generating process (DGP). The purpose of this section is not to describe the bootstrap approach but rather to demonstrate how well bootstrap performs in a given applied setting.

To generate bias-corrected efficiency scores and confidence intervals, we use the package rDEA version 1.2-5, developed by Simm and Besstremyannaya (2016). Alternatively, one could use the FEAR package developed by Paul Wilson. Details of FEAR are described in Wilson (2008). To illustrate the bootstrap approach, we only use the *provision* data of Appendix A Table A.2 and apply it to the R script in Table 3.16. The instructions also include generation of technical efficiency scores for comparisons purpose. The technical efficiency scores (TE), bias-corrected efficiency scores ($\widehat{\widehat{\theta}}$) and confidence intervals are shown in Table 3.17.

TABLE 3.18 R script for bias-corrected efficiency scores and confidence intervals for cost efficiency model (Simar and Wilson, 1998).

```
library(rDEA)

## load data
data <- read.table(file="C:/Airline/Delivery.txt", header=T,sep="\t")
attach(data)

## y refers to outputs, x refers to inputs
Y <- cbind(y1,y2)
X <- cbind(x1,x2,x3)

##Note: If costmin is used, then need a matrix called W <- cbind(w1,...)

W <- cbind(w1,w2,w3)          As cost-minimization require input prices, the new
                             instructions include the line W <- cbind(w1,w2,w3).

firms=1:16

##Technical efficiency score
ci_naive = dea(XREF=X, YREF=Y, X=X[firms,], Y=Y[firms,], W=W[firms,],
model="costmin", RTS="constant")

## Bias-corrected DEA scores and confidence intervals for bias-corrected DEA scores

ci_robust = dea.robust(X=X[firms,], Y=Y[firms,], W=W[firms,], model="costmin",
RTS="constant", B=2000, alpha=0.05, bw_mult=0.014)

data.frame(dmu,ci_naive$ gammaOpt,ci_robust$ gamma_hat_hat,ci_robust$bias,ci_robust$
gamma_ci_low,ci_robust$ gamma_ci_high)
```

The results in Table 3.17 reveal the sensitivity of the efficiency scores largely attributed to sampling variability. As such, one has to be mindful when making relative comparisons of the performances of airlines based on the TE scores instead of using the bias-corrected efficiency scores. As noted in Simar and Wilson (2011), a larger sample size would provide more consistent estimates of confidence intervals as the sample size increases.

3.6.2 Bootstrap cost-efficiency

If one wishes to obtain bias-corrected efficiency scores and confidence intervals for cost-efficiency analysis, then the following instruction shown in Table 3.18 can be used.

As our data for the '*provision*' model do not include any unit price, we therefore use the '*delivery*' model and the corresponding data (Appendix A Table A.3) to illustrate. Consequently, we also have to include $W = W$[firms,] in the R script and change the model to 'costmin'. The results therefrom are presented in Table 3.19.

TABLE 3.19 Efficiency scores, bias-corrected efficiency scores and confidence intervals for cost efficiency model (bandwidth $h = 0.014$).

3.6.3 Hypothesis test for returns to scale

As DEA is a nonparametric method, it relies on convexity assumptions. This raises the question on its underlying technology on whether it exhibits CRS or VRS. In our aforementioned analyses, we assumed CRS but may need to validate the choice of returns to scale. To test this, we apply Simar and Wilson's (2002) returns to scale bootstrapping statistics test Eq. (4.6) (see Simar and Wilson's (2002) for details on test Eq. 4.6). We use the package rDEA version 1.2-5, developed by Simm and Besstremyannaya (2016) to run Simar and Wilson's (2002) returns to scale bootstrapping statistics test and provide the script in Table 3.20.

To formally test whether the technology set T from our sample used in the *provision* model exhibits CRS, we express our null and alternative hypothesis as follows:

Ho: T is CRS
Ha: T is VRS

From Table 3.20, 'rts_input$pvalue' provides the P-value and 'rts_input$H0reject' returns a TRUE or FALSE response. A FALSE response indicates that the P-value (.373) is greater than alpha = 0.05. Thus, the null hypothesis (CRS) is not rejected in favour of the alternative (VRS).

We observe from Table 3.21 that the returns to scale bootstrapping statistics test for input-oriented DEA returned a FALSE response indicating that the P-value (.2855) is greater than alpha = 0.05. Thus, the null hypothesis (CRS) is not rejected in favour of the alternative (VRS).

TABLE 3.20 R script for Simar and Wilson (2002) returns to scale bootstrapping statistics test for input-oriented DEA for 'provision' model.

```
library(rDEA)

## load data
data <- read.table(file="C:/Airline/Provision.txt", header=T,sep="\t")
attach(data)

## y refers to outputs, x refers to inputs
Y <- cbind(y1)
X <- cbind(x1,x2)

firms = 1:16

## Testing the null hypothesis of constant returns-to-scale vs an alternative of variable
returns-to-scale
## Model may be "input" for input-oriented DEA, "output" for output-oriented DEA or
"costmin" for cost-minimisation.

rts_input=rts.test(X=X, Y=Y, W=NULL, model="input", H0="constant",
        bw="silverman", B=2000, alpha=0.05)

## If P-value is less than alpha=0.05, reject Ho for Ha. Note that a return value of FALSE in
H0reject indicates a "do not reject", TRUE otherwise.

rts_input$pvalue
rts_input$H0reject
rts_input$H0level
```

TABLE 3.21 Results from Simar and Wilson (2002) returns to scale bootstrapping statistics test for input-oriented DEA for 'provision' model.

TABLE 3.22 R script for Simar and Wilson (2002) returns to scale bootstrapping statistics test for input-oriented DEA and cost minimization for 'delivery' model.

```
library(rDEA)

## load data
data <- read.table(file="C:/Airline/Delivery_cost.txt", header=T,sep="\t")
attach(data)

## y refers to outputs, x refers to inputs
Y <- cbind(y1,y2)
X <- cbind(x1,x2,x3)
W <- cbind(w1,w2,w3)

firms = 1:16

## Testing the null hypothesis of constant returns-to-scale vs an alternative of variable
returns-to-scale
## Model may be "input" for input-oriented DEA, "output" for output-oriented DEA or
"costmin" for cost-minimisation.

rts_input=rts.test(X=X, Y=Y, W=NULL, model="input", H0="constant",
        bw="silverman", B=2000, alpha=0.05)
rts_costmin=rts.test(X=X, Y=Y, W=W, model="costmin", H0="constant",
        bw="silverman", B=2000, alpha=0.05)

## If P-value is less than alpha=0.05, reject Ho for Ha. Note that a return value of FALSE in
H0reject indicates a "do not reject", TRUE otherwise.

rts_input$pvalue
rts_input$H0reject
rts_input$H0level

rts_costmin$pvalue
rts_costmin$H0reject
rts_costmin$H0level
```

We also formally test whether the technology set T from our sample used in the *delivery* model exhibits CRS. Both technical efficiency and cost minimization were considered using R script shown in Table 3.22 and results shown in Table 3.23.

We observe from Table 3.23 that the returns to scale bootstrapping statistics test for input-oriented DEA returned a FALSE response indicating that the P-value (.236) is greater than alpha = 0.05. Thus, the null hypothesis (CRS) is not rejected in favour of the alternative (VRS). For the returns to scale bootstrapping statistics test for cost minimization model, a FALSE response indicates that the P-value (.7505) is greater than alpha = 0.05. Thus, the null hypothesis (CRS) is not rejected in favour of the alternative (VRS).

TABLE 3.23 Results from Simar and Wilson (2002) returns to scale bootstrapping statistics test for input-oriented DEA and cost minimization for 'delivery' model.

```
RStudio

File  Edit  Code  View  Plots  Session  Build  Debug  Profile  Tools  Help

Console   Terminal
C:/Program Files/RStudio/NDEA/

> ## Testing the null hypothesis of constant returns-to-scale vs an alternative of variable returns-to-scale
> ## Model may be "input" for input-oriented DEA, "output" for output-oriented DEA or "costmin" for cost-minimisation.
>
> rts_input=rts.test(X=X, Y=Y, W=NULL, model="input", H0="constant",
+                    bw="silverman", B=2000, alpha=0.05)
> rts_costmin=rts.test(X=X, Y=Y, W=W, model="costmin", H0="constant",
+                    bw="silverman", B=2000, alpha=0.05)
>
> ## If P-value is less than alpha=0.05, reject Ho for Ha. Note that a return value of FALSE in HOreject indicates a
   "do not reject", TRUE otherwise.
>
> rts_input$pvalue
[1] 0.236
> rts_input$HOreject
[1] FALSE
> rts_input$HOlevel
[1] 0.9638572
>
> rts_costmin$pvalue
[1] 0.7505
> rts_costmin$HOreject
[1] FALSE
> rts_costmin$HOlevel
[1] 0.9346147
>
> |
```

3.7 Conclusion

In this chapter, we introduced the DEA models CRS and VRS under input orientation, and allocative efficiency (cost minimization and revenue maximization) under CRS input orientation. We also briefly introduced the airline data drawn from WATS (2017). Consequently, a comprehensive analysis of airline inputs and outputs was discussed to ensure that readers recognize the importance of aligning data and the production model. We used R script to generate the CRS and VRS input-oriented efficiency scores, cost and revenue efficiency, and their respective radial and slack movements. We described what radial movement and slack movement meant for airlines that were inefficient and how efficiency can be achieved. Scale efficiency was also introduced to demonstrate the difference between CRS and VRS DEA models, and the significance of scale of operations and, with the use of NIRS, how RTS for an airline can be determined. Cost and revenue efficiency provided readers an extension of DEA analysis, which showed that airlines that are technically efficient may not necessarily be cost or revenue efficient. By including price information, cost and revenue efficiency provides additional information in the form of allocative efficiency, which is useful for business decision-making. We also introduced bootstrapping analysis to provide readers the significance of statistical inference when dealing with nonparametric methods and hypothesis testing for RTS.

Appendix A

Table A.1.

TABLE A.1 List of sample airlines.

DMU	Airline
1	Air Canada
2	ANA All Nippon Airways
3	American Airlines
4	British Airways
5	Delta Air Lines
6	Emirates
7	Garuda Indonesia
8	KLM
9	Lufthansa
10	Malaysia Airlines
11	Qantas
12	SAS Scandinavian Airlines
13	Singapore Airlines
14	TAM
15	Thai Airways
16	United Airlines

TABLE A.2 Data set for 'provision'.

DMU	Number of maintenance employees $x1$	Assets (property, plant and equipment) ('000 US$) $x2$	Available tonne kilometres (thousands) $y1$
Air Canada	2293	7,217,121	13,028,613
ANA All Nippon Airways	2591	14,651,828	14,683,532
American Airlines	11,123	26,310,000	34,707,729
British Airways	4624	19,279,420	21,401,581

TABLE A.2 Data set for 'provision'.—cont'd

DMU	Number of maintenance employees $x1$	Assets (property, plant and equipment) ('000 US$) $x2$	Available tonne kilometres (thousands) $y1$
Delta Air Lines	6628	23,357,000	27,292,425
Emirates	3457	20,837,627	27,369,447
Garuda Indonesia	102	4,736,127	2,834,184
KLM	4850	6,706,203	15,090,771
Lufthansa	19,796	31,867,956	27,007,957
Malaysia Airlines	3762	3,953,020	7,292,543
Qantas	6074	15,118,143	17,368,244
SAS Scandinavian Airlines	2047	2,954,620	4,152,670
Singapore Airlines	438	22,323,127	21,286,125
TAM	2789	8,314,066	7,840,248
Thai Airways	4620	33,144,669	10,441,041
United Airlines	4897	12,195,000	29,065,589

Source: Data for $x1$ and $x2$ are drawn from World Air Transport Statistics eDataSeries—61st Edition (2017). Data for $x2$ are drawn from each airlines' annual report and were deflated into US$ using World Bank's purchasing power parities from https://databank.worldbank.org/source/world-development-indicators.

Data set for 'provision' in text format.

```
Provision.txt - Notepad                          —    □    ×

File  Edit  Format  View  Help
dmu      x1       x2       y1
Air Canada       2293     7217121  13028613
All Nippon Airways (ANA)    2591    14651828         14683532
American Airlines    11123    26310000          34707729
British Airways    4624     19279420          21401581
Delta Air Lines    6628     23357000          27292425
Emirates  3457    20837627         27369447
Garuda Indonesia   102      4736127  2834184
KLM       4850     6706203  15090771
Lufthansa 19796    31867956         27007957
Malaysia Airlines  3762     3953020  7292543
Qantas    6074     15118143         17368244
SAS Scandinavian Airlines    2047    2954620  4152670
Singapore Airlines   438     22323127         21286125
TAM       2789     8314066  7840248
Thai Airways       4620     33144669         10441041
United Airlines    4897     12195000         29065589
```

TABLE A.3 Data set for 'delivery', 2009

DMU	Flight personnel x1	Available tonne kilometres (thousands) x2	Fuel burn (tonnes) x3	Passenger tonne kilometres performed (thousands) y1	Freight and mail tonne kilometres performed (thousands) y2	Flight personnel (US$) w1	Unit price for available tonne kilometres w2	Unit price of fuel (US$ per tonne) w3	Unit price of airfare p1	Unit price of freight and mail p2
Air Canada	8352	13,028,613	3,060,770.35	6,420,786	1,157,081	94,457.62	0.55394	673.42	1.10186	0.25755
All Nippon Airways (ANA)	6479	14,683,532	2,556,513.78	4,286,268	2,059,289	138,131.49	0.99784	348.24	1.71224	0.36933
American Airlines	23,102	34,707,729	8,654,892.94	17,866,791	2,417,898	102,330.13	0.75804	535.78	0.84162	0.23905
British Airways	16,563	21,401,581	5,304,411.47	10,079,586	4,438,214	45,438.50	0.90084	560.85	0.97544	0.17456
Delta Air Lines	17,408	27,292,425	7,349,946.47	14,571,329	1,671,083	136,609.73	0.85581	662.02	1.63382	0.47155
Emirates	13,153	27,369,447	4,717,271.61	11,276,662	6,531,110	108,950.92	0.76135	439.39	1.41748	0.46842
Garuda Indonesia	2187	2,834,184	676,346.53	1,514,745	282,129	47,916.74	1.67107	599.05	2.80117	0.98895
KLM	8101	15,090,771	3,027,818.18	7,347,192	4,093,466	84,606.91	0.44439	439.88	0.78197	0.36809

Airline										
Lufthansa	33,288	27,007,957	5,759,785.56	12,398,774	6,928,900	51,714.30	1.17995	459.97	1.22640	0.31198
Malaysia Airlines	5231	7,292,543	1,606,904.02	2,997,171	2,072,022	49,769.84	0.54206	513.20	1.64251	0.23571
Qantas	12,156	17,368,244	3,156,052.26	9,945,797	2,623,457	69,634.35	0.87045	663.49	0.78885	0.22447
SAS Scandinavian Airlines	4046	4,152,670	1,108,178.33	2,304,528	344,994	129,762.76	0.71150	831.00	1.39210	0.32857
Singapore Airlines	9467	21,286,125	3,513,668.99	7,733,939	6,559,460	105,225.91	1.04872	582.84	1.05075	0.16921
TAM	6810	7,840,248	2,015,096.39	3,935,997	155,797	173,373.99	1.06043	579.18	1.60112	4.64549
Thai Airways	7374	10,441,041	2,417,856.19	4,725,671	2,157,255	300,183.24	1.72806	527.86	2.40507	0.72576
United Airlines	18,460	29,065,589	7,647,835.29	14,645,900	2,340,509	212,296.86	0.41957	566.57	0.81320	0.22901

Source: All data, except for fuel burn, are drawn from World Air Transport Statistics eDataSeries—61st Edition (2017) and derived therefrom. Fuel burn is drawn from RDC Aviation. All unit prices, except for fuel, are derived via ratios and described in Section 3.2. Unit price of fuel is drawn from airline annual reports.

Data set for 'delivery' in text format.

Delivery.txt - Notepad

File Edit Format View Help

dmu	x1	x2	x3	y1	y2	w1	w2	w3	p1	p2
Air Canada	8352	13028613	9060770.35	6420786	1157081	97887.80	0.53449	853.69	1.16778	0.34057
ANA All Nippon Airways	6479	14683532	2556513.78	4286268	2059289	139928.95	0.89984	662.65	1.81816	0.39664
American Airlines	23102	34707729	8654892.94	17866791	2417898	93830.19	0.69987	976.10	0.94811	0.29851
British Airways	16568	21401581	5804411.47	10079586	4438214	45428.55	0.86588	494.87	1.07212	0.19948
Delta Air Lines	17408	27292425	7349946.47	14571329	1671083	91065.91	0.76854	973.02	1.28073	0.38608
Emirates	13153	27369447	4717271.61	11276662	6531110	82588.86	0.62620	634.96	1.88767	0.45538
Garuda Indonesia	2187	2834184	676346.53	1514745	282129	38761.60	1.86997	1000.86	3.38405	1.22988
KLM	8101	15090771	3027818.18	7347192	4093466	76697.63	0.40701	537.10	0.85201	0.31021
Lufthansa	33288	27007957	5759785.56	12398774		6928900	53484.77	0.97956	561.62	1.39844	0.36721
Malaysia Airlines	5231	7292543	1606904.02	2997171	2072022	44876.90	0.37161	1334.31	1.82720	0.29420
Qantas	12156	17368244	3156052.26	9945797	2623457	73410.63	0.94432	1187.68	0.85934	0.20921
SAS Scandinavian Airlines	4046	4152670	1108178.33	2304528	344994	233980.79	0.58062	1120.00	1.40462	0.30131
Singapore Airlines	9467	21286125	3513668.99	7733939	6559460	110040.80	0.89695	869.50	1.20947	0.20141
TAM	6810	7840248	2015096.39	3935997	155797	160751.07	1.87500	731.30	2.09930	5.39973
Thai Airways	7374	10441041	2417856.19	4725671	2157255	345365.56	1.60191	760.26	2.73236	0.91421
United Airlines	18460	29065589	7647835.29	14645900	2840509	220884.36	0.37934	696.48	0.95669	0.30447

References

Assaf, A.G., Josiassen, A., 2011. The operational performance of UK airlines: 2002-2007. Journal of Economic Studies 38, 5—16.

Assaf, A.G., Josiassen, A., 2012. European vs. U.S. airlines: performance comparison in a dynamic market. Tourism Management 33, 317—326.

Baltagi, B.H., Griffin, J.M., Rich, D.P., 1995. Airline deregulation: the cost pieces of the puzzle. International Economic Review 36, 245—259.

Banker, R.D., Charnes, A., Cooper, W.W., 1984. Some models for estimating technical and scale inefficiencies in data envelopment analysis. Management Science 30, 1078—1092.

Banker, R.D., 1984. Estimating most productive scale size using data envelopment analysis. European Journal of Operational Research 17 (1), 35—44.

Banker, R.D., Charnes, A., Cooper, W.W., Swarts, J., Thomas, D., 1989. An introduction to data envelopment analysis with some of its models and their uses. Research in Government and Nonprofit Accounting 5, 125—163.

Barbot, C., Costa, A., Sochirca, E., 2008. Airlines performance in the new market context: a comparative productivity and efficiency analysis. Journal of Air Transport Management 14, 270—274.

Barla, P., Perelman, S., 1989. Technical efficiency in airlines under regulated and deregulated environments. Annals of Public and Cooperative Economics 60, 61—80.

Barros, C.P., Peypoch, N., 2009. An evaluation of European airlines' operational performance. International Journal of Production Economics 122, 525—533.

Bhadra, D., 2009. Race to the bottom or swimming upstream: performance analysis of US airlines. Journal of Air Transport Management 15, 227—235.

Bogetoft, P., Otto, L., 2011. Benchmarking with DEA, SFA, and R. Springer, New York.

Bogetoft, P., Otto, L., 2015. Benchmark and Frontier Analysis Using DEA and SFA: Package 'Benchmarking'. CRAN. Version 0.26. https://cran.r-project.org/web/packages/Benchmarking/Benchmarking.pdf.

Boussofiane, A., Dyson, R.G., Thanassoulis, E., 1991. Applied data envelopment analysis. European Journal of Operational Research 52, 1—15.

Bowlin, W.F., 1998. Measuring performance: an introduction to data envelopment analysis (DEA). Journal of Cost Analysis 7, 3—27.

Charnes, A., Cooper, W.W., Rhodes, E., 1978. Measuring the efficiency of decision making units. European Journal of Operational Research 2 (6), 429–444.

Coelli, T., Rao, D.S.P., Battese, G.E., 1998. An Introduction to Efficiency and Productivity Analysis. Kluwer Academic Publishers, Boston.

Coelli, T., Perelman, S., Romano, E., 1999. Accounting for environmental influences in stochastic frontier models: with application to international airlines. Journal of Productivity Analysis 11 (3), 251–273.

Coelli, T., Rao, D.S.P., O'Donnell, C.J., Battese, G.E., 2005. An Introduction to Efficiency and Productivity Analysis, second ed. Springer, Boston.

Coll-Serrano, V., Bolos, V., Suarez, R.B., 2020. Conventional and fuzzy data envelopment analysis: R package 'deaR'. version 1.3. CRAN. https://cran.r-project.org/web/packages/deaR/deaR.pdf.

Cook, W.D., Tone, K., Zhu, J., 2014. Data envelopment analysis: prior to choosing a model. Omega 44, 1–4. https://doi.org/10.1016/j.omega.2013.09.004.

Cornwell, C., Schmidt, P., Sickles, R.C., 1990. Production frontiers with cross sectional and time-series variation in efficiency levels. Journal of Econometrics 46, 185–200.

Dakpo, K.H., Desjeux, Y., Latruffe, L., 2018. Indices of Productivity Using Data Envelopment Analysis (DEA). Package 'productivity'. Version 1.1.0. https://cran.r-project.org/web/packages/productivity/productivity.pdf.

Dyson, R.G., Allen, R., Camanho, A.S., Podinovski, V.V., Sarrico, C.S., Shale, E.A., 2001. Pitfalls and protocols in DEA. European Journal of Operational Research 132, 245–259.

Farrell, M.J., 1957. The measurement of productive efficiency. Journal of the Royal Statistical Society: Series A 120 (3), 253–290.

Førsund, F.R., 2015. Economic Perspectives on DEA, Memorandum, No. 10/2015. University of Oslo, Department of Economics, Oslo.

Friedman, L., Sinuany-Stern, Z., 1998. Combining ranking scales and selecting variables in the DEA context: the case of industrial branches. Computers & Operations Research 25 (9), 781–791.

Golany, B., Roll, Y., 1989. An application procedure for DEA. Omega 17, 237–250.

Good, D., Röller, L.H., Sickles, R.C., 1995. Airline efficiency differences between Europe and the US: implications for the pace of EC integration and domestic regulation. European Journal of Operational Research 80, 508–518.

Greer, M.R., 2008. Nothing focuses the mind on productivity quite like the fear of liquidation: changes in airline productivity in the United States, 2000–2004. Transportation Research Part A: Policy and Practice 42, 414–426.

Kuosmanen, T., Cherchye, L., Sipiläinen, T., 2006. The law of one price in data envelopment analysis: restricting weight flexibility across firms. European Journal of Operational Research 170, 735–757.

Lee, C.Y., Johnson, A.L., 2011. Two-dimensional efficiency decomposition to measure the demand effect in productivity analysis. European Journal of Operational Research 216, 84–593.

Lee, B., Worthington, A., 2014. Technical efficiency of mainstream airlines and low-cost carriers: new evidence using bootstrap data envelopment analysis truncated regression. Journal of Air Transport Management 38, 15–20.

Lee, C.Y., Johnson, A.L., 2012. Two-dimensional efficiency decomposition to measure the demand effect in productivity analysis. European Journal of Operational Research 216 (3), 584–593.

Lim, D.-J. 2021. Distance Measure Based Judgment and Learning. Package 'DJL'. Version 3.7. https://cran.rstudio.com/web/packages/DJL/index.html.

Lozano, S., Gutiérrez, E., 2014. A slacks-based network DEA efficiency analysis of European airlines. Transportation Planning and Technology 37, 623–637.

Lu, W.-M., Wang, W.-K., Hung, S.-W., Lu, E.-T., 2012. The effects of corporate governance on airline performance: production and marketing efficiency perspectives. Transportation Research Part E: Logistics and Transportation Review 48 (2), 529–544.

Nolan, J., Ritchie, P., Rowcroft, J., 2014. International mergers and acquisitions in the airline industry. In: Peoples, J. (Ed.), The Economics of International Airline Transport: Advances in Airline Economics Volume 4. Emerald Group Publishing, Bingley, UK, pp. 127–150.

Oum, T.H., Yu, C., 1995. A productivity comparison of the world's major airlines. Journal of Air Transport Management 2, 181–195.

Oum, T.H., Fu, X.-W., Yu, C., 2005. New evidences on airline efficiency and yields: a comparative analysis of major North American air carriers and its implications. Transport Policy 12, 153–164.

Panayiotis, G.M., Belegri-Roboli, A., Karlaftis, M., Marinos, T., 2009. International air transportation carriers: evidence from SFA and DEA technical efficiency results (1991–2000). European Journal of Transport and Infrastructure Research 9 (4), 347–362.

Sarkis, J., 2007. Preparing your data for DEA. In: Zhu, J., Cook, W.D. (Eds.), Modeling Data Irregularities and Structural Complexities in Data Envelopment Analysis. Springer, New York, pp. 305–320.

Schefczyk, M., 1993. Operational performance of airlines: an extension of traditional measurement paradigms. Strategic Management Journal 14 (4), 301–317.

Scheraga, C.A., 2004. Operational efficiency versus financial mobility in the global airline industry: a data envelopment and Tobit analysis. Transportation Research Part A 38, 383–404.

Schmidt, P., Sickles, R.C., 1984. Production frontiers and panel data. Journal of Business & Economic Statistics 2, 367–374.

Sengupta, J.K., 1999. A dynamic efficiency model using data envelopment analysis. International Journal of Production Economics 62 (3), 209–218.

Shott, T., Lim, D.J. 2015. TFDEA (Technology Forecasting using DEA). Package 'TFDEA'. Version 0.9.8. https://mran.microsoft.com/snapshot/2014-11-17/web/packages/TFDEA/index.html.

Simar, L., Wilson, P.W., 1998. Sensitivity analysis of efficiency scores: how to bootstrap in nonparametric frontier models. Management Science 44, 49–61.

Simar, L., Wilson, P.W., 2011. Performance of the bootstrap for DEA estimators and iterating the principle. In: Cooper, W.W., Seiford, L.M., Zhu, J. (Eds.), Handbook on Data Envelopment Analysis, second ed. Springer, Boston, pp. 241–271.

Simar, L., Wilson, P., 2002. Non-parametric tests of returns to scale. European Journal of Operational Research 139 (1), 115–132.

Simm, J., Besstremyannaya, G., 2016. Robust Data Envelopment Analysis for R: Package 'rDEA' Version 1.2-5. CRAN. https://cran.r-project.org/web/packages/rDEA/rDEA.pdf.

Vasigh, B., Fleming, K., 2005. A total factor productivity based structure for tactical cluster assessment: empirical investigation in the airline industry. Journal of Air Transportation 10, 3–19.

Wilson, P., 2008. FEAR: a software package for frontier efficiency analysis with R. Socio-Economic Planning Sciences 42, 247–254.

Zhu, J., 2011. Airlines performance via two-stage network DEA approach. Journal of CENTRUM Cathedra: Business and Economics Research Journal 4 (2), 260–269. https://doi.org/10.7835/jcc-berj2011-0063.

Chapter 4

Measuring airline productivity change

4.1 Introduction

In the previous chapter, we measured airline efficiency based on cross-sectional data. But if the focus of a study is to compare airline performance over a period of time, one would employ a productivity index approach. In this chapter, we use panel data to allow for measurement of productivity change via frontier estimation methods. Some of the index methods that we will cover in this chapter include Malmquist productivity index (MPI), Hicks—Moorsteen productivity index (HMPI), Lowe productivity index (LPI) and Färe—Primont productivity index (FPI). This chapter also recognizes the pitfalls faced by some of these index models because they fail to satisfy key axioms. Nonetheless, these index models still play an important role in the development of productivity index models and are discussed in this chapter.

To demonstrate the various productivity indices, we focus on the *delivery* model described in Chapter 3. The data for the *delivery* model are the same as Chapter 3; only this time, we focus on panel data. The data comprise three inputs and two outputs, and their respective unit prices for years 2008—2011 are shown in Appendix B Table B.1. Inputs are flight personnel, available tonne kilometres, and fuel burn. Outputs are passenger tonne kilometres performed (PTKP) and freight and mail tonne kilometres performed (FTKP). It is important to note that most values reported in annual reports are in national currencies and do not take into account inflation. As such, unit prices and values are deflated using purchasing power parities drawn from World Bank's World Development Indicators (https://databank.worldbank.org/source/world-development-indicators#).

The chapter is divided into seven sections. Following the introduction in Section 4.1, we describe the MPI in Section 4.2. Section 4.3 describes the HMPI. Section 4.4 describes the LPI. Section 4.5 describes the FPI. Section 4.6 compares the results between the aforementioned productivity models. This chapter concludes with brief comments in Section 4.7.

Productivity and Efficiency Measurement of Airlines. https://doi.org/10.1016/B978-0-12-812696-7.00005-2

4.2 Malmquist productivity index

The MPI of Caves et al. (1982), widely known as CCD Malmquist, builds on the idea of quantity indexes as ratios of distance functions first elucidated in Malmquist (1953). CCD Malmquist measures productivity change between two data points using distance functions based on DEA-like linear programming. Färe et al. (1994) extended the CCD Malmquist by taking the geometric mean of two CCD MPI between period t (base period) and $t + 1$ (next period):

$$\text{MPI}^I_{t+1}\left(y_{t+1}, x_{t+1}, y_t, x_t\right) = \left[\frac{D^I_t\left(y_{t+1}, x_{t+1}\right)}{D^I_t(y_t, x_t)} \times \frac{D^I_{t+1}\left(y_{t+1}, x_{t+1}\right)}{D^I_{t+1}(y_t, x_t)}\right]^{\frac{1}{2}} \quad (4.1)$$

where D^I is the input distance function (under input orientation as reasoned out in Section 3.6.3) and x and y are input and output bundles, respectively.[1] MPI greater than unity indicates positive productivity growth between the two periods. An equivalent way of writing this index is

$$\text{MPI}^I_{t+1}\left(y_{t+1}, x_{t+1}, y_t, x_t\right) = \left[\frac{D^I_{t+1}\left(y_{t+1}, x_{t+1}\right)}{D^I_t(y_t, x_t)}\right] \times \left[\frac{D^I_t\left(y_{t+1}, x_{t+1}\right)}{D^I_{t+1}\left(y_{t+1}, x_{t+1}\right)} \times \frac{D^I_t(y_t, x_t)}{D^I_{t+1}(y_t, x_t)}\right]^{\frac{1}{2}}$$

$$(4.2)$$

which is a decomposition of efficiency change (EFFCH) comprised of $\left[\frac{D^I_{t+1}\left(y_{t+1}, x_{t+1}\right)}{D^I_t(y_t, x_t)}\right]$ and technical change (TECH) comprised of $\left[\frac{D^I_t\left(y_{t+1}, x_{t+1}\right)}{D^I_{t+1}\left(y_{t+1}, x_{t+1}\right)} \times \right.$

$\left.\frac{D^I_t(y_t, x_t)}{D^I_{t+1}(y_t, x_t)}\right]^{\frac{1}{2}}$. Hence, we have

$$\text{MPI} = \text{EFFCH} \times \text{TECH}$$

TECH is described as a shift in the production frontier. EFFCH is the movement of DMUs either towards or away from the production frontier. EFFCH in Eq. (4.2) can be decomposed into scale efficiency change and pure technical efficiency change components using CRS and VRS DEA frontiers. The scale efficiency change (SCALECH) and pure efficiency change (PUREEFFCH) are expressed as follows:

$$\text{EFFCH} = \text{SCALECH} \times \text{PUREEFFCH}$$

---

1. As noted in Boussemart et al. (2003) that the choice of orientation makes no difference under CRS, although this issue has been controversial (see Färe et al., 1998).

whereby

$$\text{SCALECH} = \frac{D_t^{I\text{(VRS)}}(y_t, x_t)}{D_t^{I\text{(CRS)}}(y_t, x_t)} \bigg/ \frac{D_{t+1}^{I\text{(VRS)}}(y_{t+1}, x_{t+1})}{D_{t+1}^{I\text{(CRS)}}(y_{t+1}, x_{t+1})} \tag{4.3}$$

captures the changes in the deviation between the VRS frontier and the CRS frontier, and the pure efficiency change component is an efficiency change component relative to the variable returns to scale technology:

$$\text{PUREEFFCH} = \frac{D_{t+1}^{I\text{(VRS)}}(y_{t+1}, x_{t+1})}{D_t^{I\text{(VRS)}}(y_t, x_t)} \tag{4.4}$$

SCALECH is the movement along the production frontier. TECH can also be decomposed into output-biased technical change (OBTC), input-biased technical change (IBTC) and magnitude of technical change (MATC) (Färe and Grosskopf, 1996).

$$\text{TECH} = \text{OBTECH} \times \text{IBTECH} \times \text{MATECH}$$

whereby

$$\text{OBTECH} = \left[\frac{D_t^I(y_{t+1}, x_{t+1})}{D_{t+1}^I(y_{t+1}, x_{t+1})} \times \frac{D_{t+1}^I(y_t, x_{t+1})}{D_t^I(y_t, x_{t+1})} \right]^{\frac{1}{2}} \tag{4.5}$$

$$\text{IBTECH} = \left[\frac{D_{t+1}^I(y_t, x_t)}{D_t^I(y_t, x_t)} \times \frac{D_t^I(y_t, x_{t+1})}{D_{t+1}^I(y_t, x_{t+1})} \right]^{\frac{1}{2}} \tag{4.6}$$

$$\text{MATECH} = \frac{D_t^I(y_t, x_t)}{D_{t+1}^I(y_t, x_t)} \tag{4.7}$$

OBTECH measures the bias in output produced (i.e. marginal rate of transformation) due to technical change. IBTECH measures the bias in input use (i.e. marginal rate of transformation) due to technical change, which is reflected by the change in the slope of isoquant. MATECH reflects the technical change effects of the parallel shift in isoquant. It is a residual technical change that captures technical change effects not reflected in input use or output produced such as innovation. The decomposition of TECH provides information regarding Hicks-neutral technical change. If IBTECH = 1 and TECH = MATECH, then Hicks-neutral technical change exist. If OBTECH = 1 and TECH = MATECH, Hicks-neutral technical change exist. Hicks-neutral technical change is a situation whereby the isoquant changes in production are parallel when the marginal rate of transformation between two inputs (outputs) is constant (Varian, 1987). If OBTEC < 1 or > 1, this indicates producing bias as discussed in Färe et al.

(2001) and Weber and Domazlicky (1999). If IBTEC <1 or >1, this indicates using/saving bias as discussed in Färe et al. (2001).[2]

4.2.1 R script for Malmquist productivity index

The R script to derive the MPI is based on 'Productivity' package version 1.1.0 developed by Dakpo et al. (2018) as shown in Table 4.1.

The data used to derive the MPI are shown in Appendix B Table B.1. The first column is the year, followed by the DMU, the three inputs ($x1$, $x2$ and $x3$) and two outputs ($y1$ and $y2$).

The 'Productivity' package version 1.1.0 also allows for assumptions regarding the nature of the production technology. The argument tech.reg refers to technical regress, and if tech.reg=TRUE, then the production model allows technical regress, and if tech.reg=FALSE, then the production model disallows technical regress. The allowance of technical regress is, however, a contentious issue. O'Donnell (2010) argued that technical change incorporates changes in the external environment such as weather and climate and that technical regress is possible, whereas Galanopoulos et al. (2011) argued that technological regress is not possible under the sequential frontier and that any adverse effects of weather impacts on technical efficiency and not on technical change. Considering the airline industry is subject to equipment failure and unforeseen events (for example, the global financial crisis and COVID-19 pandemic), which would reduce airline performance, it is thus more meaningful to use

TABLE 4.1 R script for Malmquist productivity index.

```
## load package 'productivity'
library(productivity)

## load data
data <- read.table(file="C:/Airline/Delivery2008_11.txt", header=T,sep="\t")
attach(data)

Malmquist <- malm(data, id.var = "dmu", time.var = "Year", x.vars = c("x1", "x2", "x3"),
y.vars = c("y1", "y2"), rts = c("crs"), orientation = c("in"),
tech.reg=TRUE)

Malmquist.change <- Changes(Malmquist)
Malmquist.change
```

> Note that regardless of rts, both CRS and VRS generate the same results.

---

2. If there was only one output, the OBTECH would equal 1 for all airlines because there is no bias and thus no marginal rate of transformation.

tech.reg=TRUE to capture the effects of these events and apply this assumption to our productivity model. Our R script used for measuring the MPI assumes a CRS technology based on the argument in Färe et al. (1994) that MPI measurement under VRS can present interpretation difficulties and at times lead to computational difficulties because the distances may not always be defined in some interperiod DEA linear programming.

4.2.2 Interpretation of results

The results for the MPI (Eq. 4.2), EFFCH, TECH, SCALECH (Eq. 4.3) PUREEFFCH (Eq. 4.4), OBTECH (Eq. 4.5), IBTECH (Eq. 4.6) and MATECH (Eq. 4.7) are presented in Table 4.2. Year.1 shown in Table 4.2 means the change from the previous year. So if year.1=2009, then it means the change from 2008 to 2009. It is worth mentioning that the R script of Table 4.1 generates the Shepherd distance functions to derive the MPI. By including the following lines[3]:

Malmquist.levels $<-$ Changes(Malmquist)
Malmquist.levels

The Shephard distance function estimates are reported in the R output as shown in Table 4.2. We can illustrate, using a numerical example, on how the MPI is computed using the generated Shephard distance functions estimates. Table 4.2 shows a sample of the Malmquist level output:

Table 4.2 shows 11 columns. Column 1 is the dmu (i.e. airline). Column 2 is Year.1 (i.e. 2009) and column 3 is Year.0 (i.e. 2008). They refer to the reference technology, which in the case of our analysis, period $t + 1$ and period t, respectively. Columns 4—11 are the names of the returned Shephard distance functions according to orientation and returns to scale. The prefix 'c'

TABLE 4.2 Sample of Malmquist levels (Shephard distance function estimates).

	dmu	Year.1	Year.0	c111i	c100i	c011i	c000i	c110i	c010i	v111i	v000i
1	Air Canada	2009	2008	1.099160	1.1267136	1.0989829	1.127134	1.0473930	1.0472239	1.087439	1.115878
2	All Nippon Airways (ANA)	2009	2008	1.370918	1.2119298	1.3901597	1.228940	1.2603681	1.2780584	1.296506	1.153521
3	American Airlines	2009	2008	1.077378	1.0614647	1.0776650	1.061438	1.0008985	1.0011655	1.000000	1.000000

---

3. Dakpo et al. (2018) report this as Malmquist levels in their R package 'productivity'.

stands for CRS. If one uses a VRS, NIRS or NDRS, then a prefix v will be shown. The suffix 'i' stands for input orientation. If one uses an output orientation, then a suffix 'o' will be shown. Each distance function displays three digits. For columns 4–11, the first digit represents the period of the reference technology (i.e. either 0 or 1). The second digit represents the period of the inputs, and the third digit represents the period of the outputs.

As the Malmquist indexes of total factor productivity are defined as ratios of the Shepherd distance functions, it is noted that these distance functions are the reciprocal to the Farrell measures of technical efficiency (Färe et al., 1992; Grosskopf, 1993; Färe and Grosskopf, 2000). So we can take the reciprocal of columns 4 to 11, to derive the Farrell technical efficiency scores shown in Table 4.3.

From Table 4.3, c111i is $D^I_{t+1}(y_{t+1}, x_{t+1})$, c100i is $D^I_{t+1}(y_t, x_t)$, c011i is $D^I_t(y_{t+1}, x_{t+1})$, c000i is $D^I_t(y_t, x_t)$, c110i is $D^I_{t+1}(y_t, x_{t+1})$, c010i is $D^I_t(y_t, x_{t+1})$, v111i is $D^{I(VRS)}_{t+1}(y_{t+1}, x_{t+1})$ and v000i is $D^{I(VRS)}_t(y_t, x_t)$.

Using Air Canada as a numerical illustration to compute its MPI, this would be

$$\text{MPI}^I_{2009} = \left[\frac{0.9099}{0.8872} \times \frac{0.9098}{0.8875} \right]^{\frac{1}{2}} = 1.02534$$

Using Eq. (4.2), the numerical examples for the decompositions of MPI into EFFC and TECH are as follows:

$$\text{EFFCH}^I_{2009} = \frac{0.9098}{0.8872} = 1.0255$$

$$\text{TECH}^I_{2009} = \left[\frac{0.9099}{0.9098} \times \frac{0.8872}{0.8875} \right]^{\frac{1}{2}} = 0.9999$$

Using Eqs. (4.3) and (4.4), the numerical examples for scale efficiency and pure efficiency change are as follows:

$$\text{SCALECH}^I_{2009} = \frac{0.8962}{0.8872} \Big/ \frac{0.9196}{0.9098} = 0.9993$$

$$\text{PUREFFCH}^I_{2009} = \frac{0.9196}{0.8962} = 1.02621$$

Finally using Eqs. (4.5)–(4.7), the OBTECH, IBTECH and MATECH numerical examples are as follows:

$$\text{OBTECH} = \left[\frac{0.9099}{0.9098} \times \frac{0.9548}{0.9549} \right]^{\frac{1}{2}} = 1.0000$$

TABLE 4.3 Sample of Malmquist levels (Farrell technical efficiency estimates).

	Dmu	Year.1	Year.0	c111i	c100i	c011i	c000i	c110i	c010i	v111i	v000i
1	Air Canada	2009	2008	0.9098	0.8875	0.9099	0.8872	0.9548	0.9549	0.9196	0.8962
2	All Nippon Airways	2009	2008	0.7294	0.8251	0.7193	0.8137	0.7934	0.7824	0.7713	0.8669
3	American Airlines	2009	2008	0.9282	0.9421	0.9279	0.9421	0.9991	0.9988	1.0000	1.0000

$$\text{IBTECH} = \left[\frac{0.8875}{0.8872} \times \frac{0.9549}{0.9548}\right]^{\frac{1}{2}} = 1.00026$$

$$\text{MATECH} = \frac{0.8872}{0.8875} = 0.99962$$

The aforementioned estimates are verified in Table 4.4.

Using Air Canada (2009−10) to illustrate, we note that its Malmquist productivity change of 1.0547 (i.e. 5.47% growth) is attributed to TECH (1.1074 or 10.74% growth), while EFFCH regressed (−4.75%). EFFCH illustrates a 'catch-up' effect, which measures how much closer to the frontier the airline is, by capturing the extent of knowledge of the technology use either from changes in improved resource allocation or reduction in organizational slack. We can further investigate Air Canada's decline in EFFCH for 2009−10 by decomposing its EFFCH into SCALECH and PUREEFFCH. Its SCALECH of 0.9662 (−3.38%) and PUREEFFCH of 0.9858 (1.42%) suggest that regress in EFFCH was mainly attributed to SCALECH. This would suggest that Air Canada's scale of operations is too big, and because PUREEFFCH relates to the efficient use of resources and improvements in best practice management, it could infer that Air Canada is unable to fully utilize all of its resources efficiently.

In terms of the existence of Hicks-neutral technical change, this is a mix-bag. In 2008−09, airlines such as ANA Nippon Air, Delta Airlines, Garuda Indonesia, SAS Scandinavian Airlines and TAM exhibit Hicks-neutral technical change. But in 2009−10, only SAS Scandinavian Airlines and TAM exhibit Hicks-neutral technical change. On average, the existence of Hicks-neutral technical change for the surveyed period appeared in two airlines— SAS Scandinavian Airlines and TAM, as shown in Table 4.4.

4.2.3 Final remark

The MPI of Caves et al. (1982) is appealing because it is decomposable into technical change and efficiency change and does not require price data to estimate the production technology. However, Grifell-Tatjé and Lovell (1995) showed that if the assumed production technology is under VRS, the MPI may not properly measure the changes in total factor productivity, and that distance functions and measurement of MPI should be calculated only under CRS. Even under CRS, the decomposition estimates can be statistically inconsistent and unreliable as argued in Ray and Desli (1997) and Wheelock and Wilson (1999). O'Donnell (2010) demonstrated that the MPI is multiplicatively incomplete, suggesting that it may be an unreliable measure of TFP change because it fails to capture productivity changes associated with changes in scope (i.e. changes in output mix and input mix). Glass and McKillop (2000) and Yoruk and Zaim (2005) noted that the MPI can also generate infeasible results.

TABLE 4.4 Malmquist productivity change and its decomposition.

RStudio

File Edit Code View Plots Session Build Debug Profile Tools Help

Console   Terminal

C:/Airline/

```
> Malmquist.change <- Changes(Malmquist)
> Malmquist.change
```

dmu	Year.1	Year.0	malmquist	effch	tech	obtech	ibtech	matech	pure.inp.effch	inp.scalech
Air Canada	2009	2008	1.0253415	1.0254499	0.9998943	1.0000000	1.0002673	0.9996271	1.0261517	0.9993161
All Nippon Airways (ANA)	2009	2008	0.8840281	0.8964361	0.9861585	1.0000000	1.0000000	0.9861585	0.8897148	1.0075544
American Airlines	2009	2008	0.9850860	0.9852048	0.9998794	1.0000000	0.9998540	1.0000255	1.0000000	0.9852048
British Airways	2009	2008	0.9999888	1.0212984	0.9791348	0.9988638	0.9999547	1.0003098	1.0332100	0.9884712
Delta Air Lines	2009	2008	1.0029478	1.0026826	1.0002644	1.0000000	1.0000000	0.9958229	1.0000000	1.0026826
Emirates	2009	2008	0.9742560	0.9742560	0.9962632	0.0221863	0.9982596	0.9621753	0.9779102	0.9779102
Garuda Indonesia	2009	2008	0.9265662	0.9629910	0.9621753	0.0227019	1.0000000	0.9662386	1.0000000	0.9629910
KLM	2009	2008	0.9799882	0.9799882	0.9799882	0.0115241	1.0000000	0.9582244	1.0000000	1.0000000
Lufthansa	2009	2008	0.9522609	0.9743059	0.9773737	0.0028052	1.0000000	0.9981073	1.0000000	0.9743059
Malaysia Airlines	2009	2008	0.9987179	0.9987179	0.9740606	1.0000000	0.0931952	0.9583380	1.0549063	1.0253139
Qantas	2009	2008	1.0305894	1.0000000	0.0505894	1.0028052	1.0000000	0.9781073	1.0000000	0.0000000
SAS Scandinavian Airlines	2009	2008	0.9817663	0.9365962	0.9621733	1.0003707	0.0931952	0.9543753	0.0466123	0.0839178
Singapore Airlines	2009	2008	1.0223511	1.0000000	0.9817663	1.0000000	0.9890620	0.9943923	0.9892591	0.9952598
TAM	2009	2008	1.0022511	1.0416512	0.9621753	1.0000000	0.9890620	0.9621753	0.0466123	0.9945509
Thai Airways	2009	2008	0.9649747	0.9642641	0.9804021	0.9992935	1.0000000	0.9810953	0.9687252	0.0118283
United Airlines	2009	2008	0.9806728	0.9801835	0.0004992	1.0000000	0.9997690	1.0007304	0.9857593	0.9662108
Air Canada	2010	2009	1.0547401	0.9524513	1.073953	1.0000000	0.9917537	1.1166031	1.0000000	0.9743059
All Nippon Airways (ANA)	2010	2009	1.1245537	1.1070717	1.0157912	0.9978907	1.0000000	1.0179283	1.1600233	1.0129272
American Airlines	2010	2009	1.0342238	0.9155204	1.1296569	0.0025519	0.9975942	1.1323812	0.9038363	1.0338855
British Airways	2010	2009	0.9951166	1.0000000	1.0613923	0.9944242	1.0000000	0.0673436	0.9068294	0.9268052
Delta Air Lines	2010	2009	1.0323896	0.9375578	1.1139230	0.0016480	0.9994999	1.1144803	1.0000000	0.9268052
Emirates	2010	2009	1.0651185	1.0225888	1.0415903	1.0016480	0.9964369	1.0435949	1.0000000	1.0225888
Garuda Indonesia	2010	2009	1.0226664	0.8020970	1.1539884	0.9929683	1.0000000	1.1621603	1.0000000	0.8020970
KLM	2010	2009	1.0772795	1.0000000	1.0226664	0.0127318	0.0140285	0.9958396	1.0000000	1.0416797
Lufthansa	2010	2009	1.0955720	1.0416797	1.0341753	0.0043885	1.0000000	0.0296567	1.0000000	0.1240647
Malaysia Airlines	2010	2009	1.0789357	0.0240647	1.0698269	0.0025519	0.9998482	0.0672658	1.0000000	0.0240647
Qantas	2010	2009	1.1789357	1.0000000	1.1789357	0.0180494	0.0264830	1.1281368	1.0000000	0.9058119
SAS Scandinavian Airlines	2010	2009	1.0526986	0.9058119	1.0381973	0.0141278	1.0000000	1.1621603	1.0000000	0.0000000
Singapore Airlines	2010	2009	1.0381973	1.0000000	1.1621603	0.0000000	0.0144188	0.0091831	1.0000000	1.0000000
TAM	2010	2009	1.0544660	0.9081931	1.1621603	0.0000000	1.0000000	1.1621603	0.9433000	0.9627828
Thai Airways	2010	2009	1.0718781	1.0191151	1.0517528	0.9897361	0.9992170	1.0626384	0.0033028	1.0157802
United Airlines	2010	2009	1.0258564	0.9233505	1.1118060	1.0000000	1.0000000	1.1126773	0.9225186	0.0009019
Air Canada	2011	2010	1.0258564	0.9666215	1.0612803	1.0000000	0.0211445	1.0612803	0.9699384	0.9965803
All Nippon Airways (ANA)	2011	2010	1.0036477	0.0134306	0.9903468	0.9946979	1.0000000	0.0190897	0.9962454	1.0172499
American Airlines	2011	2010	0.9985135	1.0000000	1.0612803	1.0000000	1.0000000	0.9958257	1.9168605	0.0246016
British Airways	2011	2010	1.0046273	0.0463773	0.0612803	1.0042308	1.0000000	0.9752803	0.0225659	1.0210735
Delta Air Lines	2011	2010	0.9780872	0.9216106	0.0612803	1.0000000	1.0000000	1.0612803	1.0000000	0.9216106
Emirates	2011	2010	0.9339626	0.9471089	0.9861196	0.9980569	1.0082611	0.9799441	1.0000000	0.9471089
Garuda Indonesia	2011	2010	1.0655356	1.0245820	1.0399711	0.9993593	0.0211445	1.0190897	1.0000000	1.0245820
KLM	2011	2010	0.9930834	1.0000000	0.9930834	0.1170188	0.0047427	0.9718559	1.0000000	1.0000000
Lufthansa	2011	2010	0.9626740	0.0023664	0.9604013	0.0220679	1.0000000	0.9584194	1.0000000	0.0023664
Malaysia Airlines	2011	2010	0.9299169	0.9771888	0.9516245	1.0093187	0.0277168	0.9428385	1.0000000	0.9771888
Qantas	2011	2010	1.0366148	1.0000000	1.0366148	1.0038314	1.0000000	1.0048082	1.0000000	1.0000000
SAS Scandinavian Airlines	2011	2010	0.9913968	0.9666528	1.0048082	1.0028136	0.0289851	1.0048082	1.0000000	0.9866528
Singapore Airlines	2011	2010	0.9771390	1.0000000	0.9771390	1.0048082	1.0000000	0.9469500	1.0000000	0.9972196
TAM	2011	2010	0.9906160	0.9858757	0.0048082	1.0000000	1.0000000	1.0048082	0.9886245	1.0000000
Thai Airways	2011	2010	0.9362706	0.9764995	0.9588029	1.0015119	1.0000000	0.9573555	0.9876161	0.9887440
United Airlines	2011	2010	0.9807607	0.9241298	1.0612803	1.0000000	1.0000000	1.0612803	0.9398031	0.9833227

>

4.3 Hicks–Moorsteen productivity index

Given the limitations and pitfalls of the MPI, O'Donnell (2010, 2012a,b) proposed and showed that the HMPI satisfies a 'multiplicatively complete' total factor productivity (TFP) index and is decomposable into technical change and various measures of efficiency change. The HMPI, introduced by Bjurek (1996), has its origins from Hicks (1961) and Moorsteen (1961) as noted by Diewert (1992). Since then, the HMPI has been further refined by Färe et al. (1996), Briec and Kerstens (2004), and O'Donnell (2010, 2012a,b). The HMPI[4] is defined as the ratio of the Malmquist output and input quantity indexes and expressed as follows:

$$\text{HMPI}_t\left(y_{t+1}, x_{t+1}, y_t, x_t\right) = \left[\frac{D_{t+1}^O\left(y_{t+1}, x_{t+1}\right)D_t^O\left(y_{t+1}, x_t\right)}{D_{t+1}^O(y_t, x_{t+1})D_t^O(y_t, x_t)}\frac{D_{t+1}^I\left(y_{t+1}, x_t\right)D_t^I(y_t, x_t)}{D_{t+1}^I\left(y_{t+1}, x_{t+1}\right)D_t^I(y_t, x_{t+1})}\right]^{\frac{1}{2}}$$

(4.8)

whereby D^O and D^I are the output and input distance functions, respectively, and all notations follow those defined in Section 4.2. These distance functions can be measured using DEA models developed by O'Donnell (2010, 2012a,b). Considering an input-oriented model, the HMPI can be decomposed into a multiplicatively complete Hicks–Moorsteen TFP for airline n in period t expressed as follows:

$$\text{HMPI}_{nt}^I = \text{TFP}_t^* \times \text{TFPE}_{nt}$$

(4.9)

whereby TFP_t^* (i.e. technical change) is the maximum TFP possible using the technology in period t. Following O'Donnell (2012b), TFPE_{nt} for airline n in period t can be illustrated in Fig. 4.1 as the ratio of TFP at the observed point A to TFP at the point of maximum productivity E (i.e. TFPE_{nt} = slope OA/slope OE). TFPE_{nt} can be decomposed into several measures of efficiency change such as input-oriented technical efficiency (ITE), input-oriented scale efficiency (ISE), input-oriented mix efficiency (IME), residual input-oriented scale efficiency (RISE) and residual mix efficiency (RME) (O'Donnell, 2012b). The measures of efficiency change are described in the following and illustrated in Fig. 4.1.

- ITE $= \frac{Y_t/X_t}{\overline{Y}_t/\overline{X}_t} = \frac{\overline{X}_t}{X_t}$ measures the difference between observed TFP and the maximum TFP possible while holding the input mix, output mix and output level fixed.

- ISE $= \frac{Y_t/\overline{X}_t}{\widetilde{Y}_t/\widetilde{X}_t}$ measures the difference between TFP at a technically efficient point and the maximum TFP possible while holding the input and output mixes fixed but allowing the levels to vary.

---

4. HMPI is the geometric mean of HMPI in period t and HMPI in period $t + 1$. See O'Donnell (2010) for details on this.

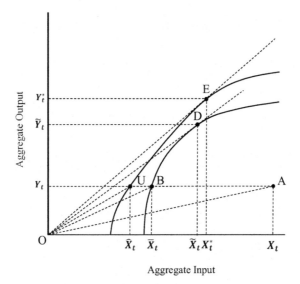

Aggregate Input

FIGURE 4.1 Input-oriented technical, scale and mix efficiency. *Source: Adapted from O'Donnell, C.J. 2012b. An aggregate quantity framework for measuring and decomposing productivity and profitability change. J. Prod. Anal. 38, 255–272.*

- IME $= \dfrac{Y_t/\overline{X}_t}{Y_t/\hat{X}_t} = \dfrac{\hat{X}_t}{\overline{X}_t}$ measures the difference between TFP at a technically efficient point on the mix-restricted frontier and the maximum TFP possible holding the output level fixed.

- RISE $= \dfrac{Y_t/\hat{X}_t}{\text{TFP}_t^*} = \dfrac{Y_t/\hat{X}_t}{Y_t^*/X_t^*}$ measures the difference between TFP at a technically and mix-efficient point and the maximum TFP possible. Potential productivity gain is achieved through economies of scale reflected by a movement from one mix-efficient point to another along the unrestricted production frontier.

- RME $= \dfrac{\widetilde{Y}_t/\widetilde{X}_t}{\text{TFP}_t^*} = \dfrac{\widetilde{Y}_t/\widetilde{X}_t}{Y_t^*/X_t^*}$ measures the difference between TFP at a point on a mix-restricted frontier and the maximum TFP possible when input and output mixes (and levels) can vary.

O'Donnell (2012b) showed that Eq. (4.9) can be expressed in various forms of decompositions:

$$\text{HMPI}_{nt}^I = \text{TFP}_t^* \times (\text{ITE}_t \times \text{IME}_t \times \text{RISE}_t) \tag{4.10}$$

$$\text{HMPI}_{nt}^I = \text{TFP}_t^* \times (\text{ITE}_t \times \text{ISE}_t \times \text{RME}_t) \tag{4.11}$$

Using Eq. (4.10) for illustration, we can compare between two airlines, n and m, operating in period t and period $t + 1$, and derive the Hicks–Moorsteen productivity change (HMPC) and its decomposition into components of technical change and efficiency change, as shown in the following:

$$\text{HMPC}^I_{nt,mt+1} = \underbrace{\left(\frac{\text{TFP}^*_{t+1}}{\text{TFP}^*_t}\right)}_{\text{technical change}} \times \underbrace{\left(\frac{\text{ITE}_{mt+1}}{\text{ITE}_{nt}} \times \frac{\text{IME}_{mt+1}}{\text{IME}_{nt}} \times \frac{\text{RISE}_{mt+1}}{\text{RISE}_{nt}}\right)}_{\text{efficiency change}} \quad (4.12)$$

From Eq. (4.12), an airline exhibits technical improvement if technical change is greater than 1.00 (i.e. $\text{TFP}^*_{t+1} > \text{TFP}^*_t$). If technical change is less than 1.00 (i.e. $\text{TFP}^*_{t+1} < \text{TFP}^*_t$), then the airline exhibits technical decline. Similarly, for the ratios within efficiency change, a value greater (less) than 1.00 indicates an improvement (decline) in efficiency from period t to period $t + 1$. O'Donnell (2012b) also showed that Eq. (4.12) can be further expressed as follows:

$$\text{TFP}^I_{nt,mt+1} = \underbrace{\left(\frac{\text{TFP}^*_{t+1}}{\text{TFP}^*_t}\right)}_{\text{technical change}} \times \underbrace{\left(\frac{\text{ITE}_{mt+1}}{\text{ITE}_{nt}} \times \frac{\text{ISME}_{mt+1}}{\text{ISME}_{nt}}\right)}_{\text{efficiency change}} \quad (4.13)$$

whereby ISME is a combined measure of scale and mix efficiency change. Alternatively, if one uses Eq. (4.11) to compare airlines n and m, operating in period t and period $t + 1$, then the decomposition of the TFP index into components of technical change and overall efficiency change would be

$$\text{TFP}^I_{nt,mt+1} = \underbrace{\left(\frac{\text{TFP}^*_{t+1}}{\text{TFP}^*_t}\right)}_{\text{technical change}} \times \underbrace{\left(\frac{\text{ITE}_{mt+1}}{\text{ITE}_{nt}} \times \frac{\text{ISE}_{mt+1}}{\text{ISE}_{nt}} \times \frac{\text{RME}_{mt+1}}{\text{RME}_{nt}}\right)}_{\text{efficiency change}} \quad (4.14)$$

4.3.1 R script for Hicks–Moorsteen productivity index

The R script to derive the HMPI is based on 'Productivity' package version 1.1.0 developed by Dakpo et al. (2018) as shown in Table 4.5.[5]

The same data set used in the measurement of MPI was used to allow us to compare the results between MPI and HMPI. From Table 4.5, Hicks1. levels generates the results for productivity and efficiency change shown in Table 4.6.

---

5. To measure the HMPI levels, change the 'Hicks1.change <− Changes(Hicks1)' to 'Hicks1. levels <− Levels(Hicks1)' as well as the corresponding 'Hicks1.change' to 'Hicks1.levels'.

TABLE 4.5 R script for Hicks–Moorsteen productivity index.

```
## load package productivity
library(productivity)

## load data
data <- read.table(file="C:/Airline/Delivery2008_11.txt ", header=T,sep="\t")
attach(data)

##Derive Hicksmoorsteen productivity index (In results, d is prefix for change)
Hicks1 <- hicksmoorsteen(data, id.var = "dmu", time.var = "Year", x.vars = c("x1", "x2",
"x3"), y.vars = c("y1", "y2"), rts = c("crs"), orientation = c("in"))

Hicks1.change <- Changes(Hicks1)
Hicks1.change
```

4.3.2 Interpretation of results

Table 4.6 presents the productivity and efficiency change results for the HMP and its decomposition into maximum productivity (MP)[6] and TFPE, and the latter into ITE, IME, RISE, ISE and RME. Note that 2008 is NA because the results refer to productivity and efficiency change (i.e. from 1 year to the next indicating that 2008 is the starting year). d for each variable refers to Δ. Productivity and efficiency change estimates greater than 1.000 indicate an improvement in the measures, and estimates less than 1.000 indicate a deterioration in these measures. ΔTech is attributed to changes in the environment, which captures the effects of technical change (i.e. technological advances) as well as the effects of government regulations and aviation policies. ΔTFPE is attributed to the size of operations and management of operations in terms of resource allocation and utilization. One could view ΔTFPE as the change in innovation efficiency as it measures the difference between observed TFP (dTFP) and the maximum TFP (dMP).

A cursory glance of Table 4.6 reveals that MP (i.e. technical change) has generally been the main contributor to TFP change while change in TFPE revealed more decline. Decomposition of TFPE in Eqs. (4.12) and (4.14) reveals that decline in efficiency change was attributed mainly to deteriorating RISE, while ISE to some extent IME contributed to improved efficiency.

Using Emirates as an example, its 2009 HMP (i.e. ΔTFP from 2008 to 2009) equals to 0.960382 indicating a decline in TFP. Its HMP is the product of ΔMP (i.e. change in maximum productivity) = technical change (1.117165) and

---

6. MP here is not to be confused with Malmquist productivity. MP refers to maximum productivity defined by Dakpo et al. (2018) in their 'Productivity' package version 1.1.0.

TABLE 4.6 Hicks–Moorsteen productivity change and its decomposition.

ΔTFPE = efficiency change (0.85966). What this suggests is that growth in HMP was mainly attributed to technical change while efficiency change declined. As noted in Eq. (4.12), ΔTFPE = ΔITE (0.97791) $\times$ ΔIME (0.962908) $\times$ ΔRISE (0.912941). Alternatively, for Eq. (4.14), ΔTFPE = ΔITE (0.97791) $\times$ ΔISE (1.00000) $\times$ ΔRME (0.879078). We also observe that ΔISE (1.00000) $\times$ ΔRME (0.879078) = ΔISME = 0.879078. Due to the decline in ΔRME and ΔRISE, what this suggests is that Emirates is operating near the unrestricted frontier but not operating at the most productive scale size. It has the potential to achieve productivity gains through economies of scale via movement from one mix-efficient point to another along the unrestricted production frontier. It would also need to adjust its input mix to operate at the optimal scale of production.

4.3.3 Final remark

The advantages of HMPI are that price data are not required for measuring productivity. HMPI does not preassume any market environment (i.e. perfect competition, regulated market or industries) or behavioral objectives (i.e. maximize output, minimize cost and maximize profit). Furthermore, there is no need to determine the orientation because HMPI combines output and input quantity indices using the Shephard's output and input distance functions, respectively (O'Donnell 2011a; María et al. 2016). However, one drawback the HMPI faces (as well as the MPI) is that it fails the transitivity test (O'Donnell, 2012a). What this suggests is that the HMPI is only appropriate for bilateral comparisons and not appropriate for multilateral comparisons.

4.4 Lowe productivity index

Productivity comparisons across multiple airlines (i.e. multilateral) and multiple periods (i.e. multitemporal) require an index that satisfies the transitivity property. Transitivity property is a property whereby a direct comparison between two periods yields the same index as an indirect comparison through a third period. An index method that satisfies the transitivity property is the LPI. Balk and Diewert (2010) and O'Donnell (2012a) attribute the LPI to Lowe (1823) because it is the ratio of the Laspeyres output quantity index to input quantity index. The Lowe quantity index uses fixed prices to weight outputs and inputs and may be from within or outside the reference period (Balk 2008; Hill 2010; O'Donnell, 2012a). LPI is appealing because it is multiplicative complete and can be exhaustively decomposed into measure of technical change and various measures of efficiency change including pure technical efficiency change, mix efficiency change and scale efficiency change.

O'Donnell (2012a) illustrates the LPI in terms of the ratios of Laspeyres indices (output and input quantity indices) as follows:

$$\text{LPI}_{nt,mt+1} = \frac{Y_{nt,mt+1}}{X_{nt,mt+1}} = \frac{p_0' y_{mt+1}}{p_0' y_{nt}} \times \frac{w_0' x_{nt}}{w_0' x_{mt+1}} \tag{4.15}$$

whereby airline m in period $t+1$ is compared with airline n in period t. Y is the aggregate output quantity index, and X is the aggregate input quantity index expressed as follows:

$$Y_{nt,mt+1} = \frac{Y(y_{mt+1})}{Y(y_{nt})} = \frac{p_0' y_{mt+1}}{p_0' y_{nt}} \tag{4.16}$$

and

$$X_{nt,mt+1} = \frac{X(x_{mt+1})}{X(x_{nt})} = \frac{w_0' x_{mt+1}}{w_0' x_{nt}} \tag{4.17}$$

LPI defined as a distance function can be derived using linear programming similar to HMPI and FPI. For detailed description of it, we direct readers to O'Donnell (2012a).

O'Donnell (2010) also showed that carefully defined price and quantity aggregates can be used to decompose profitability change (a measure of value change) into the TT index (a measure of price change) and a multiplicatively complete TFP index (a measure of quantity change) analogous to Eqs. (4.12)–(4.14). The LPI is a revenue–cost ratio. Let us denote R_{nt} and C_{nt} as the revenue and cost for airline n in period t; the bundle of output and input quantities as y and x, respectively; and the associated output and input prices as p and w, respectively. Then the profitability expression for airline n in period t can be written as

$$\text{PROF}_{it} = \frac{R_{nt}}{C_{nt}} = \frac{y_{nt} p_{nt}}{x_{nt} w_{nt}} \tag{4.18}$$

If we compare the profitability of airline n in period t with airline m in period $t + 1$, then Eq. (4.18) can be written as

$$
\begin{aligned}
\text{PROF}_{mt+1,\,nt} &= \frac{\text{PROF}_{mt+1}}{\text{PROF}_{nt}} = \frac{y_{mt+1} p_{mt+1}}{x_{mt+1} w_{mt+1}} \bigg/ \frac{y_{nt} p_{nt}}{x_{nt} w_{nt}} \\[2mm]
&= \left(\frac{y_{mt+1} p_{mt+1}}{x_{mt+1} w_{mt+1}} \right) \left(\frac{x_{nt} w_{nt}}{y_{nt} p_{nt}} \right) \\[2mm]
&= \left(\frac{y_{mt+1}}{y_{nt}} \right) \left(\frac{p_{mt+1}}{p_{nt}} \right) \left(\frac{x_{nt}}{x_{mt+1}} \right) \left(\frac{w_{nt}}{w_{mt+1}} \right) \\[2mm]
&= \underbrace{\left(\frac{y_{mt+1}}{y_{nt}} \right) \bigg/ \left(\frac{x_{mt+1}}{x_{nt}} \right)}_{\substack{\text{TFP} \\ \text{index}}} \times \underbrace{\left(\frac{p_{mt+1}}{p_{nt}} \right)}_{\substack{\text{Output} \\ \text{price} \\ \text{index}}} \bigg/ \underbrace{\left(\frac{w_{mt+1}}{w_{nt}} \right)}_{\substack{\text{Input} \\ \text{price} \\ \text{index}}}
\end{aligned} \tag{4.19}
$$

which is decomposed into a TFP index, output price index and input price index as described in O'Donnell (2010, 2011a). The TFP index component in

Eq. (4.19) is equivalent to the $\frac{TFP_{mt+1}}{TFP_{nt}}$ in Eq. (4.16), and the ratio of the output price index and input price index is the terms-of-trade index (TT).[7]

The decomposition of Eq. (4.19) is useful because it can be used to determine sources of profitability. As noted in Grifell-Tatjé and Lovell (1999); and O'Donnell (2011a), if the output price index changes at the same rate as the input price index, then PROF is attributed to TFP change. If the output change is the same as the input change, then PROF is attributed to price change.

4.4.1 R script for Lowe productivity index

The R script to derive the LPI is based on 'Productivity' package version 1.1.0 developed by Dakpo et al. (2018) as shown in Table 4.7, and the results therefrom are presented in Table 4.8.

TABLE 4.7 R script for Lowe productivity index.

```
##Lowe Productivity

## load package productivity
library(productivity)

## load data
data <- read.table(file="C:/Airline/Delivery2008_11.txt", header=T,sep="\t")
attach(data)

Lowe <- lowe(data, id.var = "dmu", time.var = "Year", x.vars = c("x1", "x2", "x3"), y.vars =
c("y1", "y2"), w.vars = c("w1", "w2", "w3"), p.vars = c("p1", "p2"), rts = "crs", orientation =
"in", by.id = NULL, by.year = 1, tech.reg=TRUE)

options(max.print=999999)

Lowe.change <- Changes(Lowe)
Lowe.change
```

4.4.2 Interpretation of Lowe productivity and profitability change results

Table 4.8 presents the productivity change (dTFP) and profitability change (dPROF) of airlines with 2008 as the benchmark year. The table also presents the decomposition of LP change into technical change (MP) and efficiency change

---

7. For details on FPI and profitability index, see O'Donnell (2010, 2011a).

TABLE 4.8 Lowe productivity and profitability change, and its decomposition (by.id = NULL and by.year = 1).

(TFPE); and the latter into ITE, IME, RISE, ISE and RME. Note that 2008 for all airlines equal to 1.000 because the results refer to productivity and efficiency change (i.e. from 1 year to the next indicating that 2008 is the base year).

We observe that in 2009, all airlines experienced a slowdown in technical change (MP). As the LP estimates of technical change over the period 2008−11 are obtained under the assumption that in any given period all airlines experience the same set of production possibilities, it is thus observed that all airlines experience the same estimated technical change in each period. As noted in Table 4.7, we assumed tech.reg = TRUE because this assumption captures any technical regress in airline performance due to environmental events that are nondiscretionary and beyond the control of airlines. As technical change is the change in the production possibilities set caused by any changes in external environment including climatic variations (O'Donnell, 2010), it is likely that the negative technical change (MP) in 2009 was attributed to the global financial crisis (GFC), which would have impacted airlines all over the world. As observed in Appendix B Table B.1, ATK ($x2$) for most airlines declined due to falling demand, suggesting reduced airline capacity thus impacting on technical change. The aftermath of the GFC reveal dTFP gaining more traction over dTT, suggesting improvements in TFPE (dTFPE) via the ability of management to adjust inputs accordingly, whereas prices such as fuel, materials and labour are subject to exogenous factors; hence, dTT was slow to recover.

4.4.3 Final remark

Balk (2008) and O'Donnell (2012a) showed that the Lowe output quantity index satisfies all economically relevant axioms from index number theory. However, in terms of price index, Balk (2008) and O'Donnell (2012a) showed that the index fails to satisfy the identity axiom, rendering the Lowe index 'less than ideal for purposes of measuring changes in prices and the TT'[8] (O'Donnell, 2012a). As the Lowe index is a price-based index, it thus means that it can only be used if price data is available.

4.5 Färe−Primont productivity index

Another index method that satisfies the transitivity property is the FPI first proposed in Färe and Primont (1995).[9] It not only satisfies the transitivity property but also satisfies the identity property and all other economically relevant axioms in the index number theory (O'Donnell, 2011a, 2014). FPI is also appealing because it is multiplicative complete and can be exhaustively decomposed into measures of technical change and various measures of

---

8. TT stands for terms of trade.
9. O'Donnell (2011b) refers it as the 'Färe−Primont' index because it can be expressed as the ratio of two indices defined by Färe and Primont (1995, p. 36, 38).

efficiency change including pure technical efficiency change, mix efficiency change and scale efficiency change. In addition, the FPI does not require price data to measure TFP.

The FPI as proposed by O'Donnell (2011a, 2014) is expressed as

$$\text{FPI}_{nt,mt+1} = \frac{\text{TFP}_{mt+1}}{\text{TFP}_{nt}} = \frac{D^O(x_0, y_{mt+1}, t_0)}{D^O(x_0, y_{nt}, t_0)} \times \frac{D^I(x_{mt+1}, y_0, t_0)}{D^I(x_{nt}, y_0, t_0)} \quad (4.20)$$

whereby D^O and D^I are the Shepherd output and input distance functions, respectively. As Eq. (4.20) is a fixed-weight index, the transitivity axiom is verified where (x_0, y_0, t_0) is the fixed vector of input and output quantities and time period, respectively.

In similar fashion to the HMPI, O'Donnell (2010, 2012a,b) demonstrated that these distance functions can be measured using DEA models and decomposed into technical change and several measures of efficiency change as shown in Eqs. (4.21)−(4.24).

$$\text{FPI}_{nt,mt+1} = \frac{\text{TFP}_{mt+1}}{\text{TFP}_{nt}} = \underbrace{\frac{\text{TFP}^*_{t+1}}{\text{TFP}^*_t}}_{\text{technical change}} \times \underbrace{\frac{\text{TFPE}_{mt+1}}{\text{TFPE}_{nt}}}_{\text{efficiency change}} \quad (4.21)$$

$$\text{FPI}_{nt,mt+1} = \underbrace{\left(\frac{\text{TFP}^*_{t+1}}{\text{TFP}^*_t}\right)}_{\text{technical change}} \times \underbrace{\left(\frac{\text{ITE}_{mt+1}}{\text{ITE}_{nt}} \times \frac{\text{ISME}_{mt+1}}{\text{ISME}_{nt}}\right)}_{\text{efficiency change}} \quad (4.22)$$

$$\text{FPI}_{nt,mt+1} = \underbrace{\left(\frac{\text{TFP}^*_{t+1}}{\text{TFP}^*_t}\right)}_{\text{technical change}} \times \underbrace{\left(\frac{\text{ITE}_{mt+1}}{\text{ITE}_{nt}} \times \frac{\text{IME}_{mt+1}}{\text{IME}_{nt}} \times \frac{\text{RISE}_{mt+1}}{\text{RISE}_{nt}}\right)}_{\text{efficiency change}} \quad (4.23)$$

$$\text{FPI}_{nt,mt+1} = \underbrace{\left(\frac{\text{TFP}^*_{t+1}}{\text{TFP}^*_t}\right)}_{\text{technical change}} \times \underbrace{\left(\frac{\text{ITE}_{mt+1}}{\text{ITE}_{nt}} \times \frac{\text{ISE}_{mt+1}}{\text{ISE}_{nt}} \times \frac{\text{RME}_{mt+1}}{\text{RME}_{nt}}\right)}_{\text{efficiency change}} \quad (4.24)$$

For the ratios within efficiency change, a value greater (less) than 1.00 indicates an improvement (decline) in efficiency from period t to period $t + 1$. Note that Eqs. (4.20)−(4.22) refer to input-oriented models to maintain consistency in the orientation throughout the book.

FPI can also be used to measure profitability change if unit prices of inputs and outputs are well defined. The Färe−Primont profitability index (FPPI) is a revenue−cost ratio and is decomposable into TT and TFP as described in Section 4.4.

4.5.1 R script for FP to measure productivity and profitability change

The R script to derive the FP productivity and profitability change is based on 'Productivity' package version 1.1.0 developed by Dakpo et al. (2018) as shown in Table 4.9.

TABLE 4.9 R script for Färe—Primont profitability index.

```
##Färe-Primont Profitability

## load package productivity
library(productivity)

## load data
data <- read.table(file="C:/Airline/Delivery2008_11.txt", header=T,sep="\t")
attach(data)

FareP1 <- fareprim(data, id.var = "dmu", time.var = "Year", x.vars =
c("x1", "x2", "x3"), y.vars = c("y1", "y2"), w.vars = c("w1", "w2",
"w3"), p.vars = c("p1", "p2"), rts = "crs", orientation = "in", by.id =
1, by.year = 1, tech.reg=TRUE)

options(max.print=999999)

FareP1.change <- Changes(FareP1)
FareP1.change
```

To measure profitability, we include the following instructions:
w.vars = c("w1", "w2", "w3"), p.vars = c("p1", "p2")

If the results return the following statement:
[reached 'max' / getOption("max.print") -- omitted 12 rows]
then the following script should be included:
options(max.print=999999)

As FPI satisfies the transitivity test, the aforementioned instructions allow one to compare airlines with a benchmark airline or compare with a benchmark year. If the instructions have the following arguments, by.id = NULL and by.year = NULL, then each airline is comparing itself for each period. If by.id = 1 and by.year = NULL, then airlines are benchmarked with dmu 1 (i.e. Air Canada as it is the first airline in the sample list) for each period. If by.id = NULL and by.year = 1, then airlines are benchmarked with itself to a specific period, which in our sample is 2008. If by.id = 1 and by.year = 1, then airlines are benchmarked with dmu 1 in 2008 (i.e. Air Canada). For purposes of presentation and interpretation of results, we present results based on by.id = NULL and by.year = 1.

4.5.2 Interpretation of Färe—Primont productivity and profitability change results

4.5.2.1 Productivity results

Table 4.10 presents the productivity of airlines with 2008 as the benchmark year. The table also presents the decomposition of FPI into technical change (MP) and efficiency change (TFPE), and the latter into ITE, IME, RISE, ISE and RME. Note that 2008 is all equal to 1.000 because the results refer to productivity and efficiency change (i.e. from 1 year to the next indicating that 2008 is the starting year). Using All Nippon Airways (ANA) as an example, its ΔTFP in 2009 was 0.983864 (i.e. -1.614%) indicating productivity decline from 2008 to 2009.

TABLE 4.10 Färe–Primont productivity change, profitability change and its decomposition (by.id = NULL and by.year = 1).

ΔTFP was mainly attributed to ΔTFPE $= 1.0108517$ with ΔMP (i.e. technical change) exhibiting a decline of 0.973302 (i.e. -2.67%). The decomposition of ΔTFPE, as shown in the following, reveals that improvements were mainly driven by scale of operations and input and output mix.

ΔISME (1.127634) $= \Delta$IME (1.114701) $\times \Delta$RISE (1.011603).
ΔISME (1.127634) $= \Delta$ISE (1.0000000) $\times \Delta$RME (1.127634).

We observe similar findings between Tables 4.8 and 4.10 that in 2009, all airlines experienced a slowdown in technical change (MP) due to the GFC and under the assumption that in any given period, airlines experience the same set of production possibilities, thus experiencing the same technical change in each period. This would be quite true during a period of downturn when airlines would not be utilizing all current technologies let alone adopting new technologies, due to falling demand.

4.5.2.2 Profitability results

The FPPI results shown in Table 4.10 reveal that most airlines suffered a decline in dPROF in 2009. The decline was attributed to decline in both dTT and dTFP. As expected during the GFC, fall in demand impacts on dTFP in terms of falling capacity utilization. Fall in dTT was due to the widening gap between input prices and output prices, whereby increases in input prices (dW) exceeded decreases in output prices (dP).

4.5.3 Final remark

The advantages of the FPI are that price data are not required for measuring productivity. But if price data are available, it allows for the measurement of sources of profitability into TFP and TT. Unlike MPI, HMPI and LPI, FPI satisfies all economically relevant axioms in the index number theory (O'Donnell, 2011a, 2014). The FPI also satisfies the transitivity property allowing for comparisons between two or more airlines, directly and indirectly, over a period of time. The results therefrom allow management of inefficient airlines to formulate policies that emulate those on the frontier to improve TFP.

4.6 A comparisons of productivity indices

In this section, we compare the productivity results from the index methods covered in this chapter. The geometric mean of productivity change, efficiency change and technical change for each airline for MPI, HMPI, LPI and FPI is presented in Table 4.11.

The MP and HMPI results for most airlines are similar, except for Lufthansa, which exhibited productivity growth under HMPI but regressed under MPI. When we observe their decompositions into EFFCH and TECH, we note more dissimilar estimates especially for TECH whereby all airlines exhibit

TABLE 4.11 Productivity change, technical change and efficiency change for MPI, HMPI, LPI and FPI, 2008–11 (geometric mean[a]).

	MPI			HMPI			LPI			FPI		
	TFP[b]	EFFCH	TECH	TFP	EFFCH	TECH	TFP	EFFCH	TECH	TFP	EFFCH	TECH
Air Canada	1.0352	0.9810	1.0553	1.0352	0.9840	1.0521	1.1177	1.1782	0.9486	1.0344	1.0395	0.9950
All Nippon Airways (ANA)	0.9993	1.0019	0.9973	0.9993	0.9883	1.0111	0.9918	1.0456	0.9486	1.0526	1.0579	0.9950
American Airlines	1.0055	0.9465	1.0623	1.0055	0.9557	1.0521	0.9891	1.0427	0.9486	1.0051	1.0101	0.9950
British Airways	0.9945	0.9879	1.0067	0.9945	0.9772	1.0177	1.0409	1.0974	0.9486	0.9928	0.9978	0.9950
Delta Air Lines	1.0042	0.9497	1.0575	1.0042	0.9545	1.0521	1.0302	1.0860	0.9486	1.0443	1.0495	0.9950
Emirates	0.9896	0.9820	1.0077	0.9861	0.8973	1.0990	1.1717	1.2352	0.9486	0.9934	0.9984	0.9950
Garuda Indonesia	0.9704	0.9250	1.0491	0.9775	0.9317	1.0491	1.1139	1.1742	0.9486	0.9165	0.9211	0.9950
KLM	0.9984	1.0000	0.9984	0.9943	0.9820	1.0126	0.9231	0.9732	0.9486	0.9832	0.9881	0.9950
Lufthansa	0.9958	1.0057	0.9902	1.0012	1.0007	1.0005	0.7888	0.8315	0.9486	1.0012	1.0062	0.9950

Malaysia Airlines	1.0058	1.0086	0.9972	1.0061	1.0044	1.0017	0.9917	1.0455	0.9486	1.0531	1.0584	0.9950
Qantas	1.0869	1.0000	1.0869	1.0926	1.0280	1.0628	1.1318	1.1931	0.9486	1.0504	1.0556	0.9950
SAS Scandinavian Airlines	1.0136	0.9749	1.0396	1.0136	0.9749	1.0396	0.9984	1.0525	0.9486	1.0150	1.0201	0.9950
Singapore Airlines	0.9918	1.0000	0.9918	0.9918	0.9917	1.0001	0.9486	1.0000	0.9486	1.0147	1.0198	0.9950
TAM	1.0157	0.9770	1.0396	1.0157	0.9770	1.0396	1.1955	1.2603	0.9486	1.1130	1.1185	0.9950
Thai Airways	0.9894	0.9931	0.9962	0.9996	0.9980	1.0016	1.0349	1.0910	0.9486	0.9977	1.0026	0.9950
United Airlines	0.9958	0.9422	1.0569	0.9958	0.9465	1.0521	0.9748	1.0276	0.9486	1.0161	1.0212	0.9950

[a] Geometric mean: Using American Airlines to illustrate the geometric mean TFP for the period 2008–11 = $(0.985086 \times 1.0342238 \times 0.9978139)^{1/3} = 1.0055$ (i.e. 0.55%).
[b] We use the term TFP here simply for the sake of comparisons with other productivity measures.
Source: Tables 4.4, 4.6, 4.8 and 4.10.

growth under HMPI but not under MPI. Our MPI results complement previous findings that both MPI and HMPI coincide under CRS, such as Bjurek (1996), Färe et al. (1996), O'Donnell (2012a,b), Kerstens and Van de Woestyne (2014), and Mizobuchi (2017).

But as noted in Färe et al. (1994), the decompositions of the MPI are only meaningful where the MPI is a true measure of TFP change, under the assumption that the technology exhibits constant returns to scale and is inversely homothetic. Cavaignac and Briec (2007) also noted that one has to be prudent to test for inverse homotheticity when using the Malmquist index. As MPI is not a true TFP measure as proven by O'Donnell (2010), that one should not rely on the MPI and its decompositions as any inference from them could lead to incorrect management decision-making. Kerstens and Van de Woestyne (2014) argued that HMPI would be the preferred TFP over MPI as 'the Malmquist productivity index maintains a TFP interpretation by approximation only when measured relative to a constant returns to scale technology' (p. 756).

When we compare LPI and FPI to HMPI, we observe that HMPI exhibits growth in TECH, whereas it exhibits decline in both LPI and FPI. A similar pattern is also reflected in EFFCH. This is the problem in using HMPI for multilateral comparisons, as the results suggest an overestimation of TECH and lead to inappropriate policies.

Comparing LPI and FPI, we observe their TFPs exhibiting some difference and some airlines having opposite outcomes. One feature of the LPI is that it uses fixed prices to weight outputs and inputs and that these price-weights in the index are fixed over time (O'Donnell, 2011a; Walden et al., 2015). As FPI is not dependent on any price data, it might suggest that the LPI is influenced by the price—weights, which may explain for the variation.

4.7 Conclusion

In this chapter, we introduce the MPI, HMPI, LPI and FPI. We illustrated the calculations of each index number and their decompositions using the R package 'Productivity' of Dakpo et al. (2018). Data used for the productivity analysis used the same inputs and outputs for the period 2008—2011 and also included price data, which are essentially implicit prices as these were indirectly derived. The literature on index number theory and applications noted that MPI and HMPI are not transitive. LPI satisfies all axioms related to index number theory but is only transitive for quantity index and not for price index. FPI is considered a true and complete TFP index not only because it satisfies all axioms related to index number theory, but it is also multiplicatively complete. Hence, the choice of index number should be determined by purpose and data availability. If one wishes to perform a bilateral comparisons, the HMPI would suffice. But if the analysis contains more than two firms, then either the LPI or FPI should be used. However, if price data are unavailable or do not exist, then the FPI should be used.

Appendix B

See Table B.1.

TABLE B.1 Data set for 'Delivery 2008_11'.

Year	DMU	Flight personnel x1	Available tonne kilometres (thousands) x2	Fuel burn (tonnes) x3	Passenger tonne kilometres performed (thousands) y1	Freight and mail tonne kilometres performed (thousands) y2	Unit price of flight personnel (US$)	Unit price of available tonne kilometres	Unit price of fuel (US$ per tonne)	Unit price of airfare	Unit price of freight and mail
2008	Air Canada	9143	13,607,675	3,237,846.8	6,738,133	1,225,028	97,887.80	0.53449	853.69	1.16778	0.34057
2008	All Nippon Airways (ANA)	6230	15,599,193	2,717,899.3	4,662,226	2,204,457	139,928.95	0.89984	662.65	1.81816	0.39664
2008	American Airlines	24,267	37,407,051	9,406,144.2	19,232,001	2,927,848	93,830.19	0.69987	976.10	0.94811	0.29851
2008	British Airways	18,390	22,398,308	5,599,237.6	10,416,066	4,809,377	45,428.55	0.86583	494.87	1.07212	0.19943
2008	Delta Air Lines	18,267	28,866,505	7,698,555.5	15,290,552	1,776,833	91,065.91	0.76854	973.02	1.28073	0.38608
2008	Emirates	13,442	23,752,982	3,881,634.9	9,608,506	6,155,580	82,588.86	0.62620	634.96	1.38767	0.45538
2008	Garuda Indonesia	2287	2,575,254	677,726.4	1,485,440	275,447	38,761.60	1.86997	1000.86	3.38405	1.22988
2008	KLM	8433	16,336,676	3,214,711.1	7,755,649	4,787,310	76,697.63	0.40701	537.10	0.85201	0.31021
2008	Lufthansa	27,376	29,350,489	5,988,763.4	12,719,586	8,474,376	53,484.77	0.97956	561.62	1.39844	0.36721

Continued

TABLE B.1 Data set for 'Delivery 2008_11'.—cont'd

Year	DMU	Flight personnel x1	Available tonne kilometres (thousands) x2	Fuel burn (tonnes) x3	Passenger tonne kilometres performed (thousands) y1	Freight and mail tonne kilometres performed (thousands) y2	Unit price of flight personnel (US$)	Unit price of available tonne kilometres	Unit price of fuel (US$ per tonne)	Unit price of airfare	Unit price of freight and mail
2008	Malaysia Airlines	5377	8,457,146	1,795,361.3	3,274,010	2,453,361	44,376.90	0.37161	1334.31	1.82720	0.29420
2008	Qantas	11,475	15,484,945	3,368,146.6	9,215,929	2,492,384	73,410.63	0.94432	1187.68	0.85934	0.20921
2008	SAS Scandinavian Airlines	4553	4,971,277	1,354,921.0	2,765,268	570,464	233,980.79	0.58062	1120.00	1.40462	0.30131
2008	Singapore Airlines	10,234	24,362,864	3,984,628.2	8,905,814	7,589,590	110,040.80	0.89696	869.50	1.20947	0.20141
2008	TAM	7069	7,034,568	1,789,061.9	3,523,594	153,848	160,751.07	1.37500	731.30	2.09930	5.39973
2008	Thai Airways	7430	10,861,259	2,555,078.5	5,053,958	2,375,369	345,365.56	1.60191	760.26	2.73236	0.91421
2008	United Airlines	19,517	31,987,031	8,315,962.8	16,031,270	2,804,892	220,884.36	0.37934	696.48	0.95669	0.30447
2009	Air Canada	8352	13,028,613	3,060,770.4	6,420,786	1,157,081	94,457.62	0.55394	673.42	1.10186	0.25755
2009	All Nippon Airways (ANA)	6479	14,683,532	2,556,513.8	4,286,268	2,059,289	138,131.49	0.99784	348.24	1.71224	0.36933

2009	American Airlines	23,102	34,707,729	8,654,892.9	17,866,791	2,417,898	102,330.13	0.75804	535.78	0.84162	0.23905
2009	British Airways	16,563	21,401,581	5,304,411.5	10,079,586	4,438,214	45,438.50	0.90084	560.85	0.97544	0.17456
2009	Delta Air Lines	17,408	27,292,425	7,349,946.5	14,571,329	1,671,083	136,609.73	0.85581	662.02	1.63382	0.47155
2009	Emirates	13,153	27,369,447	4,717,271.6	11,276,662	6,531,110	108,950.92	0.76135	439.39	1.41748	0.46842
2009	Garuda Indonesia	2187	2,834,184	676,346.5	1,514,745	282,129	47,916.74	1.67107	599.05	2.80117	0.98895
2009	KLM	8101	15,090,771	3,027,818.2	7,347,192	4,093,466	84,606.91	0.44439	439.88	0.78197	0.36809
2009	Lufthansa	33,288	27,007,957	5,759,785.6	12,398,774	6,928,900	51,714.30	1.17995	459.97	1.22640	0.31198
2009	Malaysia Airlines	5231	7,292,543	1,606,904.0	2,997,171	2,072,022	49,769.84	0.54206	513.20	1.64251	0.23571
2009	Qantas	12,156	17,368,244	3,156,052.3	9,945,797	2,623,457	69,634.35	0.87045	663.49	0.78885	0.22447
2009	SAS Scandinavian Airlines	4046	4,152,670	1,108,178.3	2,304,528	344,994	129,762.76	0.71150	831.00	1.39210	0.32857
2009	Singapore Airlines	9467	21,286,125	3,513,669.0	7,733,939	6,559,460	105,225.91	1.04872	582.84	1.05075	0.16921
2009	TAM	6810	7,840,248	2,015,096.4	3,935,997	155,797	173,373.99	1.06043	579.18	1.60112	4.64549
2009	Thai Airways	7374	10,441,041	2,417,856.2	4,725,671	2,157,255	300,183.24	1.72806	527.86	2.40507	0.72576
2009	United Airlines	18,460	29,065,589	7,647,835.3	14,645,900	2,340,509	212,296.86	0.41957	566.57	0.81320	0.22901

Continued

TABLE B.1 Data set for 'Delivery 2008_11'—cont'd

Year	DMU	Flight personnel x1	Available tonne kilometres (thousands) x2	Fuel burn (tonnes) x3	Passenger tonne kilometres performed (thousands) y1	Freight and mail tonne kilometres performed (thousands) y2	Unit price of flight personnel (US$)	Unit price of available tonne kilometres	Unit price of fuel (US$ per tonne)	Unit price of airfare	Unit price of freight and mail
2010	Air Canada	8569	13,958,344	3,219,901.2	7,005,989	1,551,113	68,698.30	0.50600	633.61	1.10148	0.24593
2010	All Nippon Airways (ANA)	6214	14,740,048	2,578,214.8	4,613,237	2,440,536	168,553.52	1.11265	616.57	1.81162	0.43471
2010	American Airlines	22,843	34,768,045	8,643,187.2	18,315,162	2,753,998	104,523.18	0.76844	689.74	0.91509	0.24401
2010	British Airways	16,656	20,844,337	5,181,557.5	9,499,861	4,578,675	49,341.43	0.95102	857.77	1.30715	0.22982
2010	Delta Air Lines	28,094	44,162,438	11,443,559.1	24,289,843	3,316,159	84,935.33	0.55411	717.45	1.12220	0.25632
2010	Emirates	14,968	31,395,942	5,433,731.8	13,686,590	8,092,699	108,127.80	0.69738	555.83	1.31834	0.47392
2010	Garuda Indonesia	2662	3,585,816	760,580.0	1,761,421	436,176	39,757.48	1.06949	771.69	2.63868	0.94445
2010	KLM	8322	14,437,621	3,072,140.6	7,598,653	3,809,579	85,212.94	0.47318	513.20	0.75499	0.39953
2010	Lufthansa	34,319	27,195,318	5,896,807.7	13,061,313	7,682,744	56,213.30	1.25818	589.15	1.32973	0.45187
2010	Malaysia Airlines	5172	7,780,030	1,621,842.3	3,440,493	2,448,825	47,889.07	0.90950	725.81	1.57045	0.23586
2010	Qantas	12,837	18,200,001	3,119,715.6	12,112,147	2,824,764	72,543.49	0.85918	788.12	0.69060	0.23062

Year	Airline										
2010	SAS Scandinavian Airlines	3739	3,986,369	1,056,254.4	2,328,821	532,331	99,895.96	0.81580	773.00	1.42442	0.31845
2010	Singapore Airlines	9857	21,535,099	3,568,899.6	8,015,679	7,087,747	116,900.79	1.01994	736.07	1.20370	0.17671
2010	TAM	8517	8,701,273	2,147,523.0	4,610,542	174,545	156,205.49	0.99191	652.49	1.49177	4.59784
2010	Thai Airways	7209	11,478,797	2,503,521.1	5,002,665	2,980,504	382,267.27	1.47500	656.16	2.37876	0.75495
2010	United Airlines	18,327	28,975,323	7,735,753.2	14,938,658	2,768,336	229,824.85	0.43465	735.92	1.59291	0.41902
2011	Air Canada	8700	14,578,014	3,327,497.9	7,297,013	1,504,098	69,029.73	0.39357	801.00	1.12826	0.25792
2011	All Nippon Airways (ANA)	6358	16,011,266	2,820,652.8	4,695,337	2,678,880	178,315.86	1.11538	807.18	1.92578	0.42113
2011	American Airlines	23,075	34,780,733	8,705,655.7	18,460,730	2,604,687	109,252.29	0.71459	957.62	0.97217	0.26990
2011	British Airways	17,686	22,743,599	5,522,346.2	10,517,793	4,781,917	51,742.37	0.88304	769.79	1.27914	0.21829
2011	Delta Air Lines	28,706	44,955,715	11,744,047.5	24,275,120	3,460,524	87,284.61	0.57156	942.23	1.24642	0.29678
2011	Emirates	17,026	34,340,999	5,949,734.3	14,597,681	8,192,133	92,069.14	0.71813	733.14	1.31795	0.45799
2011	Garuda Indonesia	3025	4,564,624	940,762.7	2,323,349	465,022	46,287.52	0.73854	1046.09	2.70046	0.85980
2011	KLM	8791	15,409,380	3,278,757.4	8,204,678	3,813,871	87,121.65	0.46712	522.43	0.83124	0.53157
2011	Lufthansa	33,603	29,944,243	6,449,860.3	14,198,118	7,963,869	57,279.02	1.20949	812.46	1.40098	0.46853

Continued

TABLE B.1 Data set for 'Delivery 2008_11':—cont'd

Year	DMU	Flight personnel x1	Available tonne kilometres (thousands) x2	Fuel burn (tonnes) x3	Passenger tonne kilometres performed (thousands) y1	Freight and mail tonne kilometres performed (thousands) y2	Unit price of flight personnel (US$)	Unit price of available tonne kilometres	Unit price of fuel (US$ per tonne)	Unit price of airfare	Unit price of freight and mail
2011	Malaysia Airlines	5231	7,773,334	1,693,754.8	3,604,071	2,067,871	47,255.78	1.00642	982.40	1.50492	0.27220
2011	Qantas	12,541	18,779,627	3,212,271.5	12,557,982	3,517,022	72,640.48	0.85323	953.08	0.66952	0.15001
2011	SAS Scandinavian Airlines	3805	4,252,768	1,136,305.7	2,463,076	518,599	97,273.24	0.79388	970.00	1.40000	0.31593
2011	Singapore Airlines	9996	22,579,442	3,843,803.6	8,133,029	7,247,286	112,609.29	0.93015	975.07	1.15744	0.18109
2011	TAM	8616	9,543,028	2,309,737.3	5,009,112	1,012,592	168,226.50	0.94126	697.21	1.35827	0.78993
2011	Thai Airways	7409	11,939,041	2,572,847.0	4,963,202	2,855,880	322,331.24	1.38795	813.42	2.51880	0.77119
2011	United Airlines	18,188	28,302,694	7,758,079.8	14,540,127	2,504,943	229,382.01	0.45649	942.23	1.78644	0.46588

References

Balk, B.M., 2008. Price and Quantity Index Numbers: Models for Measuring Aggregate Change and Difference. Cambridge University Press, New York, p. 300.

Balk, B.M., Erwin Diewert, W., 2010. The Lowe consumer price index and its substitution bias. In: Diewert, W.E., Balk, B.M., Fixler, D., Fox, K.J., Nakamura, A.O. (Eds.), Price and Productivity Measurement: Volume 6—Index Number Theory. Trafford Press, pp. 187—197 (Chapter 8).

Bjurek, H., 1996. The Malmquist total factor productivity index. The Scandinavian Journal of Economics 98 (2), 303—313.

Boussemart, J.-P., Briec, W., Kerstens, K., Poutineau, J.-C., 2003. Luenberger and Malmquist productivity indices: theoretical comparisons and empirical illustration. Bulletin of Economic Research 55 (4), 391—405.

Briec, W., Kerstens, K., 2004. A Luenberger-Hicks-Moorsteen productivity indicator: its relation to the Hicks-Moorsteen productivity index and the Luenberger productivity indicator. Economic Theory 23 (4), 925—939.

Cavaignac, L., Briec, W. (2007). Comment: testing for inversve homotheticity: a nonparametric approach. The Japanese Economic Review, 58, 524—31.

Caves, D.W., Christensen, L.R., Diewert, W.E., 1982. The economic theory of index numbers and the measurement of input, output, and productivity. Econometrica 50 (6), 1393—1414.

Dakpo, K.H., Desjeux, Y., Latruffe, L., 2018. Indices of Productivity Using Data Envelopment Analysis (DEA). Package 'productivity'. Version 1.1.0. https://cran.r-project.org/web/packages/productivity/productivity.pdf.

Diewert, W.E., 1992. Fisher ideal output, input, and productivity indexes revisited. Journal of Productivity Analysis 3, 211—248.

Färe, R., Grosskopf, S., 1996. Intertemporal Production Frontiers: With Dynamic DEA. Kluwer Academic, Boston, MA.

Färe, R., Grosskopf, S., 2000. Theory and application of directional distance functions. Journal of Productivity Analysis 13, 93—103.

Färe, R., Primont, D., 1995. Multi-output Production and Duality: Theory and Applications. Kluwer, Boston.

Färe, R., Grosskopf, S., Lindgren, B., Roos, P., 1992. Productivity changes in Swedish pharamacies 1980—1989: a non-parametric Malmquist approach. Journal of Productivity Analysis 3, 85—101.

Färe, R., Grosskopf, S., Norris, M., Zhang, Z., 1994. Productivity growth, technical progress, and efficiency changes in industrialised countries. The American Economic Review 84, 66—83.

Färe, R., Grosskopf, S., Roos, P., 1996. On two definitions of productivity. Economics Letters 53 (3), 269—274.

Färe, R., Grosskopf, S., Roos, P., 1998. Malmquist productivity indexes: a survey of theory and practice. In: Färe, R., Grosskopf, S., Russell, R.R. (Eds.), Index Numbers: Essays in Honour of Sten Malmquist. Kluwer Academic Publishers, Boston/London/Dordrecht.

Färe, R., Grosskopf, S., Lee, W.F., 2001. Productivity and technical change: the case of Taiwan. Applied Economics 33 (15), 1911—1925.

Galanopoulos, K., Surry, Y., Mattas, K., 2011. Agricultural productivity growth in the Euro-Med region: is there evidence of convergence? Outlook on Agriculture 40 (1), 29—37. https://doi.org/10.5367/oa.2011.0026.

Glass, J.C., McKillop, D.G., 2000. A post deregulation analysis of the sources of productivity growth in UK building societies. The Manchester School 68 (3), 360–385.

Grifell-Tatjé, E., Lovell, C.A.K., 1995. A note on the Malmquist productivity index. Economics Letters 47 (2), 169–175.

Grifell-Tatjé, E., Lovell, C.A.K., 1999. Profits and productivity. Management Science 45 (9), 1177–1193.

Grosskopf, S., 1993. Efficiency and productivity. In: Fried, H.O., Lovell, C.A.K., Schmidt, S.S. (Eds.), The Measurement of Productive Efficiency: Techniques and Applications. Oxford University Press, New York, pp. 160–194.

Hicks, J., 1961. Measurement of Capital in Relation to the Measurement of Other Economic Aggregates. Macmillan, London.

Hill, P., 2010. Lowe indices. In: Diewert, W.E., Balk, B.M., Fixler, D., Fox, K.J., Nakamura, A.O. (Eds.), Price and Productivity Measurement: Volume 6—Index Number Theory. Trafford Press, pp. 197–216 (Chapter 9).

Kerstens, K., Van de Woestyne, I., 2014. Comparing Malmquist and Hicks–Moorsteen productivity indices: exploring the impact of unbalanced vs. balanced panel data. European Journal of Operational Research 233, 749–758.

Malmquist, S., 1953. Index numbers and indifference surfaces. Trabajos de Estadística 4, 209–242.

María, M.S., Ramon, S.G., Francesc, H.S., 2016. Development and application of the Hicks-Moorsteen productivity index for the total factor productivity assessment of wastewater treatment plants. Journal of Cleaner Production 112, 3116–3123.

Mizobuchi, H., 2017. Productivity indexes under Hicks neutral technical change. Journal of Productivity Analysis 48, 63–68.

Moorsteen, R., 1961. On measuring productive potential and relative efficiency. Quarterly Journal of Economics 75 (3), 151–167.

O'Donnell, C.J., 2011a. The Sources of Productivity Change in the Manufacturing Sectors of the U.S. Economy. Centre for Efficiency and Productivity Analysis, Working Papers WP07/2011. University of Queensland, Queensland. https://economics.uq.edu.au/files/5199/WP072011.pdf.

O'Donnell, C.J., 2011b. Econometric Estimation of Distance Functions and Associated Measures of Productivity and Efficiency Change. University of Queensland. Centre for Efficiency and Productivity Analysis Working Papers WP01/2011. https://economics.uq.edu.au/files/5217/WP012011.pdf.

O'Donnell, C.J., 2012b. An aggregate quantity framework for measuring and decomposing productivity and profitability change. Journal of Productivity Analysis 38, 255–272.

O'Donnell, C.J., 2010. Measuring and decomposing agricultural productivity and profitability change. The Australian Journal of Agricultural and Resource Economics 54 (4), 527–560.

O'Donnell, C.J., 2012a. Nonparametric estimates of the components of productivity and profitability change in U.S. Agriculture. American Journal of Agricultural Economics 94 (4), 873–890.

O'Donnell, C.J., 2014. Econometric estimation of distance functions and associated measures of productivity and efficiency change. Journal of Productivity Analysis 41 (2), 187–200.

Ray, S.C., Desli, E., 1997. Productivity growth, technical progress, and efficiency change in industrialized countries: comment. The American Economic Review 87 (50), 1033–1039.

Varian, H., 1987. Intermediate Microeconomics: A Modern Approach. W.W. Norton & Co, New York.

Walden, J., Fissel, B., Squires, D., Vestergaard, N., 2015. Productivity change in commercial fisheries: an introduction to the special issue. Marine Policy 62, 289–293.

Weber, W.L., Domazlicky, B.R., 1999. Total factor productivity growth in manufacturing: a regional approach. Using Linear Programming" Regional Science and Urban Economics 29, 105–122.

Wheelock, D.C., Wilson, P.W., 1999. Technical progress, inefficiency and productivity change in U.S banking 1984–1993. Journal of Money, Credit, and Banking 31 (2), 212–234.

Yoruk, B.K., Zaim, O., 2005. Productivity growth in OECD countries: a comparison with Malmquist indices. Journal of Comparative Economics 33 (2), 401–420.

Chapter 5

DEA variants in measuring airline performance

5.1 Introduction

In Chapter 3, we covered the basics of DEA. In this chapter, we introduce variations of DEA models that have been developed over the years. Some of these models include metafrontier (MF) DEA, slacks-based measure (SBM), superefficiency DEA, potential gains DEA and directional distance function (DDF).

The chapter is divided into five sections. Following the introduction in Section 5.1, we describe the MF DEA model in Section 5.2. Section 5.3 describes the SBM of Tone (2001). Section 5.4 describes some superefficiency DEA models. Section 5.5 describes the potential gains DEA model. Section 5.6 describes the DDF. This chapter concludes with some brief comments in Section 5.7.

5.2 Metafrontier DEA

If one wishes to compare airlines of different business models (i.e. mainstream airlines vs low-cost airlines or between regions), one could use the MF model. The concept of MF was first developed by Battese and Rao (2002) and refined in Battese et al. (2004) for a stochastic frontier analysis. O'Donnell et al. (2008) extended the MF using a DEA model and may be expressed in terms of output orientation or input orientation. MF is defined as the boundary of an unrestricted technology set. It is the output(input) metadistance function, which gives the maximum(minimum) amount by which a firm can radially expand(contract) its output(input) vector, given an input(output) vector. As reasoned in Chapter 3, Section 3.5.1, we consider an input orientation. The input orientation MF model is formally expressed as

$$D(x, y) = \max\{\rho : \rho > 0; (x / \rho) \; \varepsilon \; P(y)\} \qquad (5.1)$$

where ρ is the maximal amount a firm can reduce its inputs while holding the quantity of outputs constant. $P(y) = \{x : (x, y) \; \varepsilon \; T^M\}$ is the input set, with given outputs, associated with this metatechnology T^M. T^M is the

Productivity and Efficiency Measurement of Airlines. https://doi.org/10.1016/B978-0-12-812696-7.00003-9

metatechnology containing all technologically feasible input−output combinations and employed to define the MF. By definition of 5.1, an observed (x, y) is technically efficient with respect to the MF if and only if $D(x, y) = 1$.

In essence, the MF is a perceived global frontier that envelops all the group frontiers. In DEA, the MF is estimated by pooling all observations for all groups, whereas group frontiers are constructed by estimating a DEA model for each group. The group frontier k, which is defined as the boundaries of restricted technology sets, is formally expressed as

$$D^k(x, y) = \max\{\rho: \rho > 0; (x/\rho) \, \varepsilon \, P^k(y)\} \tag{5.2}$$

where $P(y) = \{x : (x, y) \, \varepsilon \, T^k\}$ and T^k is the subtechnology set that firms in group k can use x to produce y.

From Eqs. (5.1) and (5.2), we can then derive the input-orientated efficiency estimates for any firm in group k with respect to the MF and express it as follows:

$$TE(x, y) = \frac{1}{D(x, y)} \tag{5.3}$$

$$TE^k(x, y) = \frac{1}{D^k(x, y)} \tag{5.4}$$

The DEA efficiency estimates are then used as distance functions between the group frontiers and the MF. The efficiency of a DMU is assessed relative to its own group frontier, and the production environment faced by the group is assessed by measuring the distance between a group frontier, $D^k(x, y)$, and the distance between the MF, $D(x, y)$. The distance between a group frontier and the MF is known as metatechnology ratio (MTR), which is expressed as

$$MTR^k(x, y) = \frac{TE(x, y)}{TE^k(x, y)} = \frac{D^k(x, y)}{D(x, y)} \tag{5.5}$$

We can illustrate the MF, group frontier and MTR in Fig. 5.1 adapted from O'Donnell et al. (2008) although we note that our illustration reflects an input orientation.

Suppose Airline A from group 2 produces at the input−output combination labelled A in Fig. 5.1. If the MF is the convex function labelled M-M', then the measure of input-orientated technical efficiency for Airline A is

$TE(x, y)$ (i.e. A relative to M-M') = 1/(0D/0B)
$TE^k(x, y)$ (i.e. A relative to group $k = 2 - 2'$) = 1/(0C/0B)

Then following Eq. (5.5),

$MTR^k(x, y) = (0B/0D)/(0B/0C) = 0C/0D$

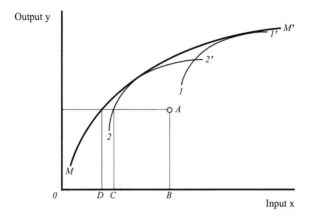

FIGURE 5.1 Technical efficiencies and metatechnology ratios (input orientation). *Source: Adapted from O'Donnell C.J., Rao, D.S.P., Battese, G.E., 2008. Metafrontier frameworks for the study of firm-level efficiencies and technology ratios. Empirical Economics 34, 231–255.*

5.2.1 R script for metafrontier

The R script for MF is shown in Table 5.1 based on the 'Benchmarking' package developed by Bogetoft and Otto (2011, 2015).

We continue our focus on the 'delivery' model and use the delivery data set of Appendix A Table 3, but only for inputs and outputs. However, due to the nature of the MF model, it is a good idea to set up the data by clusters, which in our case is by continent. We do this by sorting the airline list starting with airlines from the Americas, followed by Asia and then Europe. We do this so that the data are sorted by continent (i.e. Asia, Europe and Americas) as shown in Table 5.2. It is important to note that the DMU number and associated airline in Table 5.2 only apply for the MF in this section.

From Table 5.2, the first five rows (or DMUs) are airlines from the Americas. Rows 6–12 are airlines from Asia. Rows 13 to 16 are airlines from Europe. We thus set up our data using Appendix A Table 3 according to the requirements of MF and present them in Appendix C Table C.1. In doing so, we can then instruct in R how the data should be read by continent (i.e. Cluster).

5.2.2 Interpretation of metafrontier results for the 'delivery' model

Tables 5.3, 5.4 and 5.5 are the technical efficiency scores, i.e. $TE^k(x, y)$ for Americas, Asia and Europe, respectively. Table 5.6 presents the MF results

TABLE 5.1 R script for Metafrontier.

```
## load Benchmarking
library(Benchmarking)

## load data (Americas)
data_americas <- (read.table(file="C:/Airline/Delivery_Meta.txt",header=T,sep="\t")
[-c(6:16),])
attach(data_americas)
```
> [-c6:16,] indicates that we are instructing R to ignore the rows from 6 to 16 in dataset Appendix C Table 1.

```
## y refers to outputs, x refers to inputs
y <- cbind(y1,y2)
x <- cbind(x1,x2,x3)

# To derive Technical efficiency
tecrs_americas<- dea(x,y,RTS="crs",ORIENTATION="in")
data.frame(dmu,tecrs_americas$eff)
detach(data_americas)

## load data (Asia)
data_asia <-
```
> [-c1:5,13:16,] indicates that we are instructing R to ignore the rows from 1 to 5, and 13 to 16 in dataset Appendix C Table 1.

```
(read.table(file="C:/Airline/Delivery_Meta.txt",header=T,sep="\t")
[-c(1:5,13:16),])
attach(data_asia)

## y refers to outputs, x refers to inputs
y <- cbind(y1,y2)
x <- cbind(x1,x2,x3)

# To derive Technical efficiency
tecrs_asia<- dea(x,y,RTS="crs",ORIENTATION="in")
data.frame(dmu,tecrs_asia$eff)
detach(data_asia)

## load data (Europe)
data_europe <-
```
> [-c1:12,] indicates that we are instructing R to ignore the rows from 1 to 12 in dataset Appendix C Table 1.

```
(read.table(file="C:/Airline/Delivery_Meta.txt",header=T,sep="\t")
[-c(1:12),])
attach(data_europe)

## y refers to outputs, x refers to inputs
y <- cbind(y1,y2)
x <- cbind(x1,x2,x3)

# To derive Technical efficiency
tecrs_europe<- dea(x,y,RTS="crs",ORIENTATION="in")
```

TABLE 5.1 R script for Metafrontier.—cont'd

```
data.frame(dmu,tecrs_europe$eff)
detach(data_europe)

## DEA Metafrontier
## To derive Metafrontier Technical efficiency

data <- read.table(file="C:/Airline/Delivery_Meta.txt", header=T,sep="\t")
attach(data)

## y refers to outputs, x refers to inputs
y <- cbind(y1,y2)
x <- cbind(x1,x2,x3)

metatecrs <- dea(x,y,RTS="crs",ORIENTATION="in")
data.frame(dmu,metatecrs$eff)

## Summary of efficiencies

tecrs_americas <-as.vector(tecrs_americas$eff)
tecrs_americas <-t(tecrs_americas)
tecrs_asia <-as.vector(tecrs_asia$eff)
tecrs_asia<-t(tecrs_asia)
tecrs_europe <-as.vector(tecrs_europe$eff)
tecrs_europe <-t(tecrs_europe)
metatecrs<-as.vector(metatecrs$eff)
metatecrs<-t(metatecrs)

data.frame(mean(tecrs_americas),mean(tecrs_asia),mean(tecrs_europe),mean(metatecrs))
data.frame(sd(tecrs_americas),sd(tecrs_asia),sd(tecrs_europe),sd(metatecrs))

## Metatechnology ratios
tedea<-cbind(tecrs_americas,tecrs_asia,tecrs_europe)
MTR<-metatecrs/tedea
MTR

## Metafrontier group
MTR_americas<-mean(metatecrs)/mean(tecrs_americas)
MTR_asia<-mean(metatecrs)/mean(tecrs_asia)
MTR_europe<-mean(metatecrs)/mean(tecrs_europe)
data.frame(MTR_americas,MTR_asia, MTR_europe)
```

$TE(x, y)$ for input orientation under CRS, which are the same as Table 3.10. Table 5.7 presents the MTR results.

Using the airlines in the Americas as an illustration, we observe that from Table 5.3, most airlines were operating efficiently (e.g. Air Canada, American Airlines, Delta Airlines and United Airlines) when measured relative to the Americas frontier. However, when we consider the MTR results in Table 5.7, the results reveal that Air Canada, American Airlines and United Airlines are now operating between 91% and 94% efficiency. This difference implies a lower MTR suggesting that the maximum output that is feasible using the American technology (and the input levels used by these airlines) is only about

TABLE 5.2 Airline by continent.

DMU by row	Airline	Continent
1	Air Canada	Americas
2	American Airlines	Americas
3	Delta Air Lines	Americas
4	TAM	Americas
5	United Airlines	Americas
6	ANA All Nippon Airways	Asia
7	Emirates	Asia
8	Garuda Indonesia	Asia
9	Malaysia Airlines	Asia
10	Qantas	Asia
11	Singapore Airlines	Asia
12	Thai Airways	Asia
13	British Airways	Europe
14	KLM	Europe
15	Lufthansa	Europe
16	SAS Scandinavian Airlines	Europe

TABLE 5.3 Technical efficiency scores (Americas).

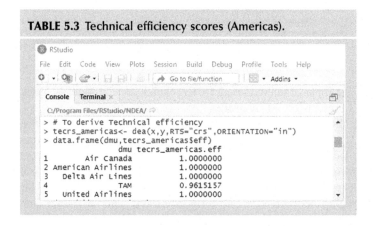

TABLE 5.4 Technical efficiency scores (Asia).

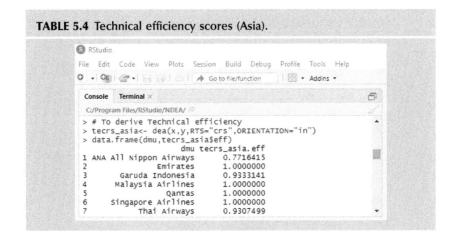

TABLE 5.5 Technical efficiency scores (Europe).

91%−94% of the output that could be achieved using the technology represented by the MF. The average MTR for airlines in the Americas (0.9423) suggests that airlines could produce 94% of the outputs that could be produced using the metatechnology (and the same inputs). The average MTR for airlines in Asia (0.9864) suggests that airlines could produce 98.6% of the outputs that could be produced using the metatechnology (and the same inputs). The average MTR for airlines in Europe (0.9642) suggests that airlines could produce 96% of the outputs that could be produced using the metatechnology (and the same inputs).

We observe that TE for airlines in Asia reveal four airlines operating efficiently. Of the four, only Qantas and Singapore Airlines are on the MF. Emirates is technically efficient at the group level but inefficient under MF.

Using Fig. 5.2 to illustrate Emirates position, as it is technically efficient in its group, it could either be at point A or B of the group frontier 2-2′. Either

TABLE 5.6 MF efficiency estimates.

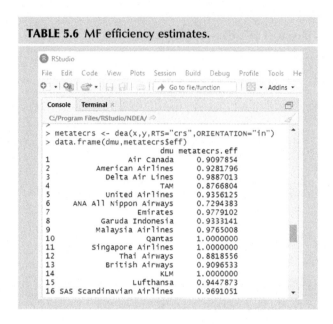

TABLE 5.7 MTR results.

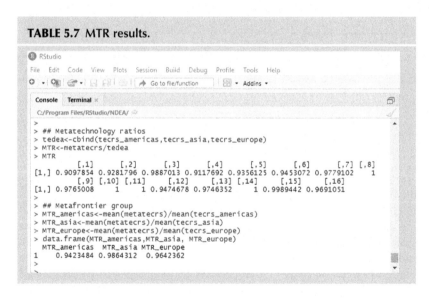

points on this group frontier will exhibit TE = 1. However, when compared with the MF, only point C is technically efficient, which makes points A and B inefficient. This inefficiency is due to the technology gap and is the distance between MF and MTR.

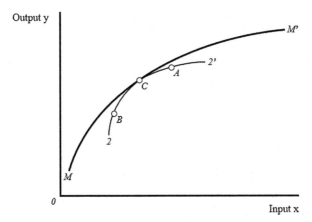

FIGURE 5.2 TE, MTR and RTE (input orientation). *Source: Adapted from O'Donnell C.J., Rao, D.S.P., Battese, G.E., 2008. Metafrontier frameworks for the study of firm-level efficiencies and technology ratios. Empirical Economics 34, 231–255.*

TABLE 5.8 R script for input-orientated SBM ('delivery' model).

```
library(deaR)

## Load data
data <- read.table(file="C:/Airline/Delivery.txt", header=T,sep="\t")
attach(data)
data_example <- read_data(data,ni = 3,no = 2)    ┐  "ni" refers to the number of inputs and "no"
                                                  ┘  refers to the number of outputs.
##Orientation = io refers to input-oriented; no = non-oriented
SBM <- model_sbmeff(data_example,orientation ="io",rts = "crs")
CCR <- model_basic(data_example,orientation = "io",rts = "crs")

data.frame(efficiencies(SBM),slacks(SBM))
data.frame(efficiencies(CCR),slacks(CCR))
```

As defined by O'Donnell (2018, p.216), "if more than one technology exists in a given period, then each measure of efficiency … can be decomposed into the product of a metatchnology ratio and a measure of residual efficiency". That is, TE = MTR × RTE, where MTR measures how well management has chosen the production technology (i.e. choosing the type of technology) and RTE (residual technical efficiency) is a measure of how well management has deployed and used the chosen technology (i.e. choosing the combination and amount of inputs). From this description, we can infer that Emirates is not operating at the optimal scale and can improve its position via better use of its inputs either by adjusting either the amount of inputs and/or combination of inputs. To determine Emirates' scale of operations, we use Table 3.11, which shows the TE for the different returns to scale and the description on Fig. 3.2. We observe that Emirates' NIRS technical efficiency = VRS technical efficiency, thus it exhibits decreasing returns to scale. This suggests that Emirates should reduce its scale of operations in accordance with the lambda weights of peers in Table 3.10.

5.3 Slacks-based measure

The SBM proposed by Tone (2001) is a nonradial model, which does not assume proportional changes of inputs and outputs. The SBM deals with slacks directly, whereas the CCR radial model does not take account of the input excesses and output shortfalls that are reflected by the nonzero slacks as described in Section 3.5.2.1.

The SBM efficiency, θ, under CRS is expressed as

$$\theta = \min_{\lambda, s^-, s^+} \frac{1 - \frac{1}{m}\sum_{i=1}^{m}\left(\frac{s_i^-}{x_{io}}\right)}{1 + \frac{1}{u}\sum_{r=1}^{u}\left(\frac{s_r^+}{y_{ro}}\right)} \tag{5.6}$$

subject to

$$x_{io} = X\lambda_j + s_i^- \quad (i = 1, \ldots, m)$$

$$y_{ro} = Y\lambda_j - s_r^+ \quad (r = 1, \ldots, u)$$

$$\lambda_j \geq 0 \ (\forall j), \qquad s_i^- \geq 0 \ (\forall i), \quad s_r^+ \geq 0 \ (\forall r)$$

where there are n DMUs, each using m inputs to produce u outputs. Vectors of inputs and outputs for DMU$_j$ using technology set T are $x_j = \left(x_{1j}, x_{2j}, \ldots, x_{mj}\right)^T$ and $y_j = \left(y_{1j}, y_{1j}, \ldots, y_{uj}\right)^T$, respectively. Thus, the input and output matrices X and Y are defined as $X = \sum_{j=1}^{n} x_{ij} = (x_1, x_2, \ldots, x_n) \in R^{m \times n}$ and $Y = \sum_{j=1}^{n} y_{rj} = (y_1, y_2, \ldots, y_n) \in R^{u \times n}$. The input and output slacks are $s^- = \left(s_1^-, s_2^-, \ldots, s_m^-\right)^T \in R^m$ and $s^+ = \left(s_1^+, s_2^+, \ldots, s_u^+\right)^T \in R^u$, respectively. $\lambda = (\lambda_1, \lambda_2, \ldots, \lambda_n)^T$ is the intensity vector. As Eq. (5.6) is in fractional form, Tone (2001) transforms it into a linear form using the Charnes–Cooper transformation (Charnes and Cooper 1962). For details of this, see Tone (2001).

Cooper et al. (2007) showed that the numerator of Eq. (5.6) is the input-oriented SBM and the denominator is the output-orientated SBM. For the former, all input slacks are zero, but output slacks may be nonzero; whereas for the latter, all output slacks are zero, but input slacks may be nonzero. When both input and output orientations are incorporated as in Eq. (5.6), all input and output slacks are zero.

5.3.1 R script for slacks-bases measure

The R script for SBM is shown in Table 5.3 based on 'deaR' package developed by Coll-Serrano et al. (2018). As the study has been using an input-orientated model, we shall maintain an input-orientated SBM model. In this case, only the numerator of Eq. (5.6) is considered.

5.3.2 Interpretation of slacks-based measure results for the 'delivery' model

Table 5.9 presents the SBM and CCR results (same as Tables 3.10 and 3.12) for comparisons purpose. The data set used is drawn from Appendix Table A Table 5.3 for the first five columns (i.e. $x1$, $x2$, $x3$, $y1$ and $y2$).

The SBM results reveal the same three efficient airlines (KLM, Qantas and Singapore Airlines) as efficient. But even when the airlines are efficient, they still exhibit negative/positive slacks. When these slacks are dealt with in the SBM, the results reveal that the same three airlines remains efficient with no slacks, while the remaining airlines' efficiency estimates are now worse than the CCR due to the existence of slacks.

TABLE 5.9 SBM and CCR ('delivery' model) results.

We also observed in Table 5.9 that the SBM absorbs the radial movements into the slacks. When we compare the radial and slack movements of British Airways described in Chapter 3 Section 3.5.4, the SBM slacks equal the standard DEA radial + slack movements.

5.4 Superefficiency DEA

In many cases, using the standard DEA results in more than one DMU with an efficiency score of 1, which occurs often with small sample sizes. To compare the relative efficiency between efficient DMUs, one has to move away from the standard DEA. Andersen and Petersen (1993) proposed the concept of superefficiency using an input-orientated CRS model. Their aim was to rank efficient DMUs under the assumption that the DMU being evaluated is excluded from the reference set. The CCR superefficiency model, using notations of Eq. (5.7), can be expressed as follows:

$$\min \theta - \varepsilon \left(\sum_{i=1}^{m} s_{io}^- + \sum_{r=1}^{u} s_{ro}^+ \right) \tag{5.7}$$

$$\text{subject to} \quad \sum_{j=1, j \neq o}^{n} x_{ij} \lambda_j + s_{io}^- = \theta x_{io}, \quad i = 1, 2, \ldots, m$$

$$\sum_{j=1, j \neq o}^{n} y_{rj} \lambda_j + s_{ro}^+ = \theta y_{ro}, \quad r = 1, 2, \ldots, u$$

$$\lambda_j, \ s_{io}^-, s_{ro}^+ \geq 0, \ j = 1, 2, \ldots, n, \ j \neq o, \ i = 1, 2, \ldots, m,$$

$$r = 1, 2, \ldots, u$$

where $\varepsilon > 0$ is the non-Archimedean infinitesimal. Eq. (5.7) allows for an efficiency score ≥ 1.

5.4.1 R script for Andersen and Petersen (1993) superefficiency DEA

The R script to generate Andersen and Petersen's superefficiency estimates is shown in Table 5.10.

5.4.2 Interpretation of Andersen and Petersen (1993) superefficiency results for the 'delivery' model

Table 5.11 presents the CRS and VRS results of Andersen and Petersen (1993) superefficiency DEA. It also includes the efficiency estimates of the standard DEA input-orientated CRS model for comparisons. The data set used is drawn from Appendix Table A Table 5.3 for the first five columns (i.e. $x1$, $x2$, $x3$, $y1$ and $y2$).

TABLE 5.10 R script for Andersen and Petersen (1993) superefficiency DEA.

```
## load TFDEA
library(TFDEA)

## load data
data <- read.table(file="C:/Airline/Delivery.txt", header=T,sep="\t")
attach(data)

## y refers to outputs, x refers to inputs
y <- cbind(y1,y2)
x <- cbind(x1,x2,x3)

# Andersen and Petersen (1993) radial DEA super-efficiency. If one wishes to view only
efficiency estimates, include $eff, otherwise exclude it to view efficiency estimates and
lambdas)

DEA_CRS<-DEA(x,y, rts="crs", orientation="input")
SDEA_CRS<-SDEA(x,y, rts="crs", orientation="input")
SDEA_VRS<-SDEA(x,y, rts="vrs", orientation="input")

data.frame(dmu, DEA_CRS$eff,SDEA_CRS$eff,SDEA_VRS$eff)
```

TABLE 5.11 Results using Andersen and Petersen (1993) superefficiency DEA.

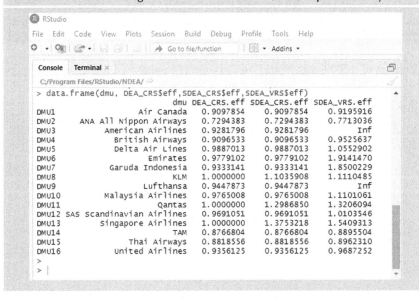

```
> data.frame(dmu, DEA_CRS$eff,SDEA_CRS$eff,SDEA_VRS$eff)
                              dmu DEA_CRS.eff SDEA_CRS.eff SDEA_VRS.eff
DMU1                    Air Canada   0.9097854    0.9097854    0.9195916
DMU2        ANA All Nippon Airways   0.7294383    0.7294383    0.7713036
DMU3             American Airlines   0.9281796    0.9281796          Inf
DMU4               British Airways   0.9096533    0.9096533    0.9525637
DMU5               Delta Air Lines   0.9887013    0.9887013    1.0552902
DMU6                      Emirates   0.9779102    0.9779102    1.9141470
DMU7              Garuda Indonesia   0.9333141    0.9333141    1.8500229
DMU8                           KLM   1.0000000    1.1035908    1.1110485
DMU9                     Lufthansa   0.9447873    0.9447873          Inf
DMU10             Malaysia Airlines   0.9765008    0.9765008    1.1101061
DMU11                        Qantas   1.0000000    1.2986850    1.3206094
DMU12    SAS Scandinavian Airlines   0.9691051    0.9691051    1.0103546
DMU13             Singapore Airlines   1.0000000    1.3753218    1.5409313
DMU14                           TAM   0.8766804    0.8766804    0.8895504
DMU15                  Thai Airways   0.8818556    0.8818556    0.8962310
DMU16               United Airlines   0.9356125    0.9356125    0.9687252
>
> |
```

We observe that the inefficient airlines have the same efficient estimates for both DEA_CRS and SDEA_CRS models. The efficient airlines under DEA_CRS remain efficient under SDEA_CRS, but their efficiency estimates are higher under SDEA_CRS suggesting the latter having more discriminatory power over the DEA_CRS model. Our finding is similar to Andersen and Petersen (1993). However, Seiford and Zhu (1999) argued that if Eq. (5.7) included the convexity constraint $\sum_{j=1}^{n} \lambda_j = 1$, $\lambda_j \geq 0$, $j = 1, 2, ..., n$, which becomes an SDEA_VRS model, it leads to infeasible solutions for certain DMUs. We note two airlines exhibiting infeasible solutions.

Since then, the issue of solving the infeasibility of superefficiency DEA under VRS has drawn considerable interest in the development of a feasible model, for example, Lovell and Rouse (2003), Chen (2004, 2005), Cook et al. (2009), Chen and Liang (2011), and Lee et al. (2011). Other variations of superefficiency models include SBM super-efficiency model such as Tone (2002) and the additive superefficiency of Du et al. (2010), the directional distance function (DDF) superefficiency model Nerlove—Luenberger (N-L) superefficiency model by Ray (2008), the modified DDF-based superefficiency model of Chen et al. (2013a), and the integer-valued additive superefficiency models of Du et al. (2012) and Chen et al. (2013b). Given the wide range of superefficiency models, we present Cook et al. (2009) modified superefficiency model and Tone (2002) Super SBM. Cook et al. (2009) showed that their model is similar to Lovell and Rouse (2003) and yields the same result as the input-orientated VRS superefficiency model. Although Cook et al. (2009) model becomes infeasible when there zero input data are present (Lee and Zhu, 2012), this situation is highly unlikely for airline performance evaluation as all input data are nonzero. We provide the superefficiency DEA models of Cook et al. (2009) and Tone (2002) in the following.

5.4.3 Cook et al. (2009) modified superefficiency DEA

Cook et al. (2009) modified superefficiency measure under VRS for an efficient DMU$_o$ is expressed as

$$\text{Min} \tau + M + \beta \tag{5.8}$$

$$\text{subject to} \quad \sum_{j=1, j \neq o}^{n} \lambda_j x_{ij} \leq (1+\tau)x_{io}, \quad i = 1, 2, ..., m$$

$$\sum_{j=1, j \neq o}^{n} \lambda_j x_{rj} \geq (1-\beta)y_{ro}, \quad r = 1, 2, ..., u$$

$$\sum_{j=1, j \neq o}^{n} \lambda_j = 1$$

$$\beta, \lambda_j \geq 0, \, j = 1, 2, ..., n, \, j \neq o$$

where M is a user-defined large positive number and is set equal to 10^5.

TABLE 5.12 R script for Cook et al. (2009) modified superefficiency DEA (CRS and VRS).

```
## load TFDEA
library(TFDEA)

## load data
data <- read.table(file="C:/Airline/Delivery.txt", header=T,sep="\t")
attach(data)

## y refers to outputs, x refers to inputs
y <- cbind(y1,y2)
x <- cbind(x1,x2,x3)

SDEA_Cook_crs<-SDEA(x,y, rts="crs", orientation="input", slack=TRUE, cook=TRUE)
data.frame(dmu,SDEA_Cook_crs)

SDEA_Cook_vrs<-SDEA(x,y, rts="vrs", orientation="input", slack=TRUE, cook=TRUE)
data.frame(dmu,SDEA_Cook_vrs)
```

5.4.4 R script for Cook et al. (2009) modified superefficiency DEA

The R script to generate Cook et al. (2009) modified superefficiency estimates is shown in Table 5.12 based on the R package "TFDEA" developed by Shott and Lim (2015). For purposes of comparisons, we provide both CRS and VRS models. The CRS model of Cook et al. (2009) is obtained by removing the constraint $\sum_{j=1}^{n}\lambda_j = 1, \ \lambda_j \geq 0, \ j = 1, 2, ..., n$ from Eq. (5.8).

5.4.5 Interpretation of Cook et al. (2009) modified superefficiency results for the 'delivery' model

In Tables 5.13 and 5.14, se.eff and se.excess refer to τ and β, respectively, as indicated in Eq. (5.3). From the information of τ and β, we observe that KLM, Qantas and Singapore Airlines have positive τ and zero β, suggesting that these airlines are superefficient in inputs. In Table 5.14, we observe more superefficient airlines under VRS. This is to be expected because CRS efficiency estimate is decomposed of 'pure' technical efficiency and scale efficiency. This suggests that apart from KLM, Qantas and Singapore Airlines, these airlines may be technically efficient but are not operating at their optimal scale. We also observe that American Airlines is superefficient in outputs but not in inputs, whereas Lufthansa is superefficient in both inputs and outputs.

TABLE 5.13 Results using Cook et al. (2009) modified superefficiency DEA (CRS).

```
> data.frame(dmu,SDEA_Cook_crs)
```

	dmu	eff	status1	status2	se.eff	se.excess	lambda.DMU1	lambda.DMU2	lambda.DMU3	lambda.DMU4	lambda.DMU5	lambda.DMU6	lambda.DMU7	lambda.DMU8
DMU1	Air Canada	0.9097854	0	0	-0.09021462	0	0	0	0	0	0	0	0	0.2834347
DMU2	ANA All Nippon Airways	0.7294383	0	0	-0.27056166	0	0	0	0	0	0	0	0	0.5833886
DMU3	American Airlines	0.9281796	0	3	-0.07182038	0	NA	NA	NA	NA	NA	NA	NA	NA
DMU4	British Airways	0.9096533	0	0	-0.09034670	0	0	0	0	0	0	0	0	0.82556335
DMU5	Delta Air Lines	0.9887013	0	3	-0.01129873	0	NA	NA	NA	NA	NA	NA	NA	NA
DMU6	Emirates	0.9779102	0	0	-0.02208978	0	0	0	0	0	0	0	0	1.0515964
DMU7	Garuda Indonesia	0.9333141	0	0	-0.06668594	0	0	0	0	0	0.07269515	0	0	0.00000000
DMU8	KLM	1.1035908	0	0	0.10359083	0	0	0	0	0	0	0	0	0.00000000
DMU9	Lufthansa	0.9447873	0	0	-0.05521272	0	0	0	0	0	0	0	0	1.6777490
DMU10	Malaysia Airlines	0.9765008	0	0	-0.02349919	0	0	0	0	0	0	0	0	0.2198326
DMU11	Qantas	1.2986850	0	0	0.29868495	0	0	0	0	0	0	0	0	1.3536868
DMU12	SAS Scandinavian Airlines	0.9691051	0	0	-0.03089488	0	0	0	0	0	0.17092261	0	0	0.00000000
DMU13	Singapore Airlines	1.3753218	0	0	0.37532177	0	0	0	0	0	0	0	0	1.3297156
DMU14	TAM	0.8766804	0	0	-0.12331964	0	0	0	0	0	0	0	0	0.00000000
DMU15	Thai Airways	0.8818556	0	0	-0.11814443	0	0	0	0	0	0	0	0	0.00000000
DMU16	United Airlines	0.9356125	0	3	-0.06438752	0	NA	NA	NA	NA	NA	NA	NA	NA

	lambda.DMU9	lambda.DMU10	lambda.DMU11	lambda.DMU12	lambda.DMU13	lambda.DMU14	lambda.DMU15	lambda.DMU16	sx.x1	sx.x2	sx.x3	sy.y1	sy.y2
DMU1	0	0	0.4361980	0	0	0	0	0	0.0000	1906947516	549791.66	NA	114749047
DMU2	0	0	0.0000000	0	0	0	0	0	0.0000	0	98424.57	0	328792368
DMU3	NA	NA	NA	NA	NA	NA	NA	NA	NA	NA	NA	NA	NA
DMU4	0	0	0.4035888	0	0	0	0.000000	0	3472.6720	NA	1051771.83	0	0
DMU5	NA	NA	NA	NA	NA	NA	NA	NA	NA	NA	NA	NA	NA
DMU6	0	0	0.1350288	0	0	0.285418836	0	0	0.0000	2474786517	150576.93	117423532	0
DMU7	0	0	0.1523000	0	0	0	0	0	189.7989	0	482536.41	0	0
DMU8	0	0	0.3299208	0	0	0.419722504	0	0	0.0000	0	329128.67	0	0
DMU9	0	0	0.0000000	0	0	0.009313509	0	0	17770.4634	0	273653.75	0	0
DMU10	0	0	0.0000000	0	0	0.178695292	0	0	1635.5033	0	329128.67	0	0
DMU11	0	0	0.0000000	0	0	0	0	0	4820.5973	2127699247	342656.43	2917813987	0
DMU12	0	0	0.2317087	0	0	0	0	0	1104.3480	0	0.00	262883892	0
DMU13	0	0	0.0000000	0	0	0	0	0	0.0000	4530780123	0.00	3963173379	0
DMU14	0	0	0.3957448	0	0	0	0	0	1159.5200	0	517604.31	882422348	0
DMU15	0	0	0.1630119	0	0	0	0	0	1098.3382	0	338390.64	0	0
DMU16	NA	NA	NA	NA	NA	NA	NA	NA	NA	NA	NA	NA	NA

TABLE 5.14 Results using Cook et al. (2009) modified superefficiency DEA (VRS).

```
RStudio
File  Edit  Code  View  Plots  Session  Build  Debug  Profile  Tools  Addins  Help

Console   Terminal ×
C:/Program Files/RStudio/NDEA/

> data.frame(dmu,SDEA_Cook_vrs)
                           dmu       eff status1 status2      se.eff   se.excess lambda.DMU1 lambda.DMU2 lambda.DMU3 lambda.DMU4   lambda.DMU5  lambda.DMU6  lambda.DMU7 lambda.DMU8
DMU1               Air Canada 0.9193916       0       0 -0.08040840  0.00000000           0           0  0.00000000 0.000000e+00 0.000000e+00  0.00000000   0.3216062   0.3130651
DMU2    ANA All Nippon Airways 0.7713036       0       3 -0.12859603          NA          NA          NA          NA           NA           NA          NA          NA          NA
DMU3         American Airlines 2.1035036       0       0 -0.12859603  0.18027250           0           0  0.00000000 3.862949e-13 0.15306175  0.00000000   0.00000000   0.2220970
DMU4           British Airways 0.9525637       0       0 -0.04743631  0.00000000           0           0  0.23124791 0.000000e+00 0.15306175  0.00000000   0.00000000   0.1222971
DMU5           Delta Air Lines 1.0553902       0       0  0.05529017  0.00000000           0           0  0.06248364 0.000000e+00 0.00000000  0.09104622   0.00000000   0.1122658
DMU6                  Emirates 1.9141470       0       0  0.91414700  0.00000000           0           0  0.00000000 0.000000e+00 0.00000000  0.97058907   0.2148020   0.0000000
DMU7          Garuda Indonesia 1.8500229       0       0  0.85002286  0.00000000           0           0  0.00000000 0.000000e+00 0.09104622  0.97058907   0.7148540   0.0000000
DMU8                       KLM 1.1110485       0       0  0.11104852  0.00000000           0           0  0.02941093 0.000000e+00 0.60594164  0.71485400   0.0000000   0.3733755
DMU9                 Lufthansa 2.1023044       0       0  0.02137576  0.07486951           0           0  0.00000000 0.000000e+00 0.60594164  0.92873124   0.9063245   0.0000000
DMU10        Malaysia Airlines 1.1101061       0       0  0.11010610  0.00000000           0           0  0.02068287 0.000000e+00 0.92873124  0.00000000   0.0000000   0.0000000
DMU11                   Qantas 1.3206094       0       0  0.32060944  0.00000000           0           0  0.00000000 0.000000e+00 0.00000000  0.00000000   0.4808502   0.4487225
DMU12 SAS Scandinavian Airlines 1.0103546      0       0  0.01035462  0.00000000           0           0  0.00000000 0.000000e+00 0.00000000  0.00000000   0.0000000   0.0000000
DMU13       Singapore Airlines 1.5409313       0       0  0.54093129  0.00000000           0           0  0.00000000 0.000000e+00 0.00000000  0.00000000   0.0000000   0.0000000
DMU14                      TAM 0.8895504       0       0 -0.11044963  0.00000000           0           0  0.00000000 6.091764e-01 0.08778485  0.00000000   0.0000000   0.0000000
DMU15             Thai Airways 0.8962310       0       0 -0.10376903  0.00000000           0           0  0.00000000
DMU16          United Airlines 0.9687252       0       0 -0.03127481  0.00000000           0           0  0.22288974

      lambda.DMU9 lambda.DMU10 lambda.DMU11 lambda.DMU12 lambda.DMU13 lambda.DMU14 lambda.DMU15 lambda.DMU16         sx.x1      sx.x2      sx.x3      sy.y1       sy.y2
DMU1   0.00000000   0.36532874   0.00000000   0.00000000   0.00000000           NA           NA   0.00000000        0.0000         NA   496240.8          0  1173598861
DMU2           NA           NA           NA           NA           NA           NA           NA           NA            NA         NA        0.0          0           0
DMU3   0.20675907   0.00000000   0.41810204   0.00000000   0.00000000           0            0   1.00000000     1953.9199 1603651685        0.0  789783000  3584491506
DMU4   0.00000000   0.00000000   0.00000000   0.00000000   0.31378890           0            0   0.65648630        0.0000          0  1147906.9          0           0
DMU5   0.00000000   0.00000000   0.00000000   0.00000000   0.41086200           0            0   0.00000000        0.0000          0   394280.9          0   884119165
DMU6   0.62372743   0.00000000   0.00000000   0.00000000   0.28513600           0            0   0.00000000   2669552567 3793675.4          0
DMU7   0.00000000   0.00000000   0.00000000   1.00000000   0.00000000           0            0   0.00000000   1090635196  143078.2          0   62865000
DMU8   0.28328981   0.00000000   0.00000000   0.00000000   0.00000000           0            0   0.00000000    20553.9471  451572.2          0
DMU9   0.00000000   0.00000000   0.00000000   0.00000000   0.00000000           0            0   0.00000000     1544.1750 1049824.7          0
DMU10  0.00000000   0.00000000   0.00000000   0.00000000   0.00000000           0            0   0.00000000     4580.8475  298464.7  290890049
DMU11  0.00000000   0.00000000   0.00000000   0.00000000   0.00000000           0            0   0.00000000      967.0437        0.0          0 2912423438
DMU12  0.09367550   0.00000000   0.00000000   0.00000000   0.00000000           0            0   0.00000000        0.0000   211018.9          0  156460068
DMU13  0.21350760   0.00000000   0.00000000   0.78649240   0.00000000           0            0   0.00000000      280.2914   622752.2 3622694531
DMU14  0.07042724   0.00000000   0.00000000   0.00000000   0.00000000           0            0   0.00000000     1065.9731   247114.8          0
DMU15  0.00000000   0.00000000   0.00000000   0.00000000   0.00000000           0            0   0.00000000        0.0000   260814.0          0   675666165
DMU16  0.08014901   0.00000000   0.00000000   0.00000000   0.00000000           0            0   0.00000000        0.0000   335090.5          0
```

5.4.6 Tone (2002) superefficiency SBM

Tone (2002) proposed a superefficiency SBM (henceforth Super SBM) model to differentiate the SBM-efficient DMUs by first identifying the efficient DMUs. Suppose that DMU_0 is SBM-efficient. Its Super SBM under CRS is calculated as the optimal objective function written as follows:

$$\theta_o^* = \min \theta_o \frac{\frac{1}{m}\sum_{i=1}^{m}\left(\frac{\overline{x}_{io}}{x_{io}}\right)}{\frac{1}{u}\sum_{r=1}^{u}\left(\frac{\overline{y}_{ro}}{y_{ro}}\right)} \quad (5.9)$$

subject to

$$\overline{x}_{io} \geq \sum_{j=1,\neq o}^{n} x_{ij}\lambda_j, \quad (i=1,2,...,m)$$

$$\overline{y}_{ro} \leq \sum_{j=1,\neq o}^{n} y_{rj}\lambda_j, \quad (r=1,2,...,u),$$

$$\overline{x}_{io} \geq x_{io} \quad (i=1,2,...,m)$$

$$\overline{y}_{ro} \leq y_{ro} \quad (r=1,2,...,u)$$

$$\lambda_j, \overline{y}_{ro} \geq 0, j=1,2,...,n, j\neq o, r=1,2,...,u$$

where the numerator of Eq. (5.9) is the input-orientated Super SBM and the denominator is the output-orientated Super SBM. As Eq. (5.1) is in fractional form and requires positive input and output values for SBM-efficient DMUs, i.e. $x_{ij} > 0$ and $y_{rj} > 0$, Tone (2002) transforms it into a linear form using the Charnes—Cooper transformation (Charnes and Cooper, 1962). For details of this, see Tone (2002). As out example in the following focusses only on input orientation, we use the numerator of Eq. (5.9).

5.4.7 R script for Tone (2002) superefficiency SBM

The R script to generate the Super SBM efficiency estimates is shown in Table 5.15 based on the R package 'deaR' developed by Coll-Serrano et al. (2018).

TABLE 5.15 R script for Super SBM.

```
## load deaR
library(deaR)

## Load data
data <- read.table(file="C:/Airline/Delivery.txt", header=T,sep="\t")
attach(data)
data_example <- read_data(data,ni = 3,no = 2)

result <- model_sbmsupereff(data_example,orientation = "io",rts = "crs",
compute_target=TRUE)

data.frame(efficiencies(result))
data.frame(round(slacks(result)$slack_input))
data.frame(targets(result)$project_input)

## To report the reference DMUs lambdas, include the following
references(result)
```

5.4.8 Interpretation of Tone (2002) super SBM results for the 'delivery' model

The results of Tone (2002) Super SBM under CRS is presented in Table 5.16. Similar to previous analysis, the results show that KLM, Qantas and Singapore Airlines are superefficient. We can illustrate how the Super SBM efficiency estimates are calculated using the projected_input information. Using KLM as an example, its superefficiency SBM estimate is $(8986.78/8101 + 16{,}541{,}048{,}926/15{,}090{,}771{,}000 + 3{,}027{,}818.2/3{,}027{,}818.2)/3 = 1.06{,}848$.

5.5 Potential gains DEA

Mergers provide opportunities for efficiency gains via economies of scale or scope (Goergen and Renneboog, 2004; Neary, 2007; Bogetoft and Wang, 2005). In addition to economies of scale and scope; Prince and Simon (2017) noted that mergers can improve service quality, share best practices and gain market power. With regards to our study period (2008−2011), several airlines had embarked on mergers and acquisitions. SAS Scandinavian Airlines acquired Air Baltic in 2008. Lufthansa acquired German Wings in 2009. Delta Air Lines merged with Northwest Airlines in 2010 (approved in October 2008 and merger completed in January 2010). United Airlines merged with Continental Airlines and began operations on November 2011. British Airways merged with Iberia in January 2011.

Even when the benefits from merging are identified, merger talks can still fall through, such as the merger talks between British Airways and Qantas in late 2008. Would such a merger bring about the benefits as anticipated? To measure this outcome, we quantify the potential gains from a merger between

TABLE 5.16 Efficiency estimates and slacks using Tone (2002) Super SBM.

```
R RStudio

File   Edit   Code   View   Plots   Session   Build   Debug   Profile   Tools   Help

O  ▾  Oₒ  ⬤ ▾   ⬜ ⬜   ⬛   ↗ Go to file/function       ▾ Addins ▾

Console   Terminal ×

C:/Program Files/RStudio/NDEA/ ⬩

> data.frame(efficiencies(result))
                            efficiencies.result.
Air Canada                             1.00000
ANA All Nippon Airways                 1.00000
American Airlines                      1.00000
British Airways                        1.00000
Delta Air Lines                        1.00000
Emirates                               1.00000
Garuda Indonesia                       1.00000
KLM                                    1.06848
Lufthansa                              1.00000
Malaysia Airlines                      1.00000
Qantas                                 1.15829
SAS Scandinavian Airlines              1.00000
Singapore Airlines                     1.29603
TAM                                    1.00000
Thai Airways                           1.00000
United Airlines                        1.00000
> data.frame(round(slacks(result)$slack_input))
                              x1         x2       x3
Air Canada                    91  307470280        0
ANA All Nippon Airways         0 5192705155   370415
American Airlines           1702 1374109775        0
British Airways             3645          0  1400581
Delta Air Lines                0          0  1927761
Emirates                       0 2273514365        0
Garuda Indonesia             269          0    38419
KLM                            0          0   191063
Lufthansa                  18698          0   843594
Malaysia Airlines           1654          0   335991
Qantas                      1190          0        0
SAS Scandinavian Airlines   1183          0   267686
Singapore Airlines             0          0        0
TAM                         1721  194246140   108451
Thai Airways                1095          0   461818
United Airlines                0  410908358   483926
> data.frame(targets(result)$project_input)
                                   x1          x2         x3
Air Canada                   8352.000 13028613000  3060770.3
ANA All Nippon Airways       6479.000 14683532000  2556513.8
American Airlines           23102.000 34707729000  8654892.9
British Airways             16563.000 21401581000  5304411.5
Delta Air Lines             17408.000 27292425000  7349946.5
Emirates                    13153.000 27369447000  4717271.6
Garuda Indonesia             2187.000  2834184000   676346.5
KLM                          8986.782 16541048926  3027818.2
Lufthansa                   33288.000 27007957000  5759785.6
Malaysia Airlines            5231.000  7292543000  1606904.0
Qantas                      12156.000 20428177859  4098717.6
SAS Scandinavian Airlines    4046.000  4152670000  1108178.3
Singapore Airlines          12981.221 24181783541  4851842.5
TAM                          6810.000  7840248000  2015096.4
Thai Airways                 7374.000 10441041000  2417856.2
United Airlines             18460.000 29065589000  7647835.3
```

British Airways and Qantas in 2008. In turn, the economic effects from this merger should then be observed in 2009, which coincidentally is our benchmark year.

To measure the potential gains from a merger, we use Bogetoft and Wang (2005) proposed method, which we term it as 'potential gains DEA' (PGDEA). We present a brief description of PGDEA in the following and recommend the avid reader to Bogetoft and Wang (2005) for details on their proposed method.

Assume the merged airline DMU^J where J is the number of merged airlines. Through direct pooling of inputs and outputs, we obtain an airlines that uses $\sum_{j \in J} x^j$ to produce $\sum_{j \in J} y^j$. Hence, Bogetoft and Wang (2005) radial input-based input orientation PGDEA from merging the J number of airlines is

$$\text{Min } \theta^J$$

$$\theta^J, \lambda$$

$$\text{subject to } -\sum_{j \in J} y_i{}^j + Y\lambda \geq 0$$

$$\theta^J \sum_{j \in J} x_i{}^j + X\lambda \geq 0$$

$$\lambda \geq 0 \tag{5.10}$$

where θ^J is the maximal proportion reduction in the aggregated inputs $\sum_{j \in J} x^j$ that allows the production of the aggregated output $\sum_{j \in J} y^j$. $\theta^J < 1$ indicates that airlines can save while $\theta^J > 1$ is costly to merge. Bogetoft and Wang (2005) expresses θ^J as

$$\theta^J = T^J \times S^J \times H^J \tag{5.11}$$

which is decomposed of learning (i.e. technical efficiency index denoted as T^J), scope or mix or harmony index (H^J) and scale or size index (S^J), and they are written as follows.

Technical efficiency index (T^J)

Technical efficiency or learning is associated with the ability to adjust to best practices. Bogetoft and Wang (2005) propose that (x^j, y^j) can be projected into $(\theta^j x^j, y^j)$, where θ^j is the standard TE estimate under input orientation for a DMU. The next step then involves using the projected plans $(\theta^j x^j, y^j)$ for basis to calculate the real merger gains:

$$\text{Min } \theta^{*J}$$

$$\theta^{*J}, \lambda$$

$$\text{subject to } -\sum_{j \in J} y_i^j + Y\lambda \geq 0$$

$$\theta^{*J}\sum_{j\in J}\theta^j x_i^j + X\lambda \geq 0$$

$$\lambda \geq 0 \tag{5.12}$$

Letting $T^J = \theta^J/\theta^{*J}$, this can be rewritten it as $\theta^J = T^J/\theta^{*J}$. T^J indicates what can be saved by individual adjustments in the different DMUs in J. If individual technical inefficiencies have been addressed, we are then left with scope or harmony effects and scale or size effects.

Scope or harmony index (H^J)

Scope or harmony index refers to the input and output mix resulting from a merger. This description of input and output mix is analogous to O'Donnell (2010). Bogetoft and Wang (2005) propose to capture the harmony gains by examining how much of the average input could have been saved in the production of the average output from the measure H^J:

$$\text{Min } H^J$$

$$H^J, \lambda$$

$$\text{subject to } \alpha\sum_{j\in J} y_i^j + Y\lambda \geq 0$$

$$H^J\alpha\sum_{j\in J}\theta^j x_i^j - X\lambda \geq 0$$

$$\lambda \geq 0 \tag{5.13}$$

where $\alpha = |J|^{-1}$ to remove any size effects. $\alpha \in [0, 1]$ is a scalar determining the size of the DMU evaluated with the harmony measure. If $H^J < 1$, it indicates a savings potential due to improved harmony, while $H^J > 1$ indicates a cost of harmonizing the inputs and outputs.

Scale or size index (S^J)

Size gains can be captured by asking how much is saved when operating at full scale rather than at α-scale. This is reflected by the measure S^J:

$$\text{Min } S^J$$

$$S^J, \lambda$$

$$\text{subject to } -\sum_{j\in J} y_i^j + Y\lambda \geq 0$$

$$S\left[H^J\sum_{j\in J}\theta^j x_i^j\right] - X\lambda \geq 0$$

$$\lambda \geq 0 \tag{5.14}$$

TABLE 5.17 R script for Bogetoft and Wang (2005) merger DEA.

```
## load Benchmarking
library(Benchmarking)

## load data
data <- read.table(file="C:/Airline/Delivery.txt",header=T,sep="\t")
attach(data)

## y refers to outputs, x refers to inputs
y <- cbind(y1,y2)
x <- cbind(x1,x2,x3)

te <- dea(x,y,RTS="crs",ORIENTATION="in")
data.frame(dmu,te$eff)

grouping <- list(c(4,11))
M <- make.merge(grouping,X=x)

em <- dea.merge(x,y,M, RTS="crs", ORIENTATION="in")
em
```

c(4,11) indicates the dmu number according "to the dataset "Delivery". DMU 4 and 11 corresponds to British Airways and Qantas, respectively.

If $S^J < 1$, it indicates that rescaling is beneficial through gains in economies of scale. If $S^J > 1$, it indicates that there are no benefits from the returns to scale property for larger DMUs and that the merger is costly.

5.5.1 R script for Bogetoft and Wang (2005) merger DEA

The R script to generate Bogetoft and Wang (2005) merger DEA is shown in Table 5.17.

5.5.2 Interpretation of PGDEA results

The results based on the PGDEA model use the 'delivery' data of Appendix Table A Table 5.3 for the first five columns (i.e. x1, x2, x3, y1 and y2). The results therefrom are presented in Table 5.18.

Using the reported indicators of Table 5.18, $learning = $Eff/$Estar corresponds to $T^J = \theta^J/\theta^{*J} = 0.950/1 = 0.950$. $Eff = $learning × $harmony × $size equates to Eq. (5.5) and equals to $0.950 × 1 × 1 = 0.950$.

From our merger analysis of British Airways and Qantas, the technical efficiency estimate of 0.95 indicates potential gains from the merger of 0.05 through the best practice. However, the estimates of harmony and size equaling 1 indicate that they are operating at the CRS part of the frontier, which suggests that a merger does not provide any gains. The individual efficiency scores of British Airways (0.9097) and Qantas (1.000) also seem to suggest that British Airways has more to gain from the merger. The

TABLE 5.18 Results using Bogetoft and Wang (2005) PGDEA.

aforementioned findings would thus support Qantas' chief executive Alan Joyce comment that in addition to other significant matters, an appropriate merger ratio would need to be resolved (Sydney Morning Herald, https://www.smh.com.au/business/qantasba-merger-off-20081218-71gf.html).

As mergers and acquisitions increase the asset size of airlines, the question of scale of operations needs to be asked. Following the analysis of Section 3.5.2.2, when we tested the scale of operations using the R script of Table 3.7 with data set Appendix B Table 1, we found all US airlines and Lufthansa exhibit DRS for all years. These are the same airlines that engaged in either mergers or acquisitions. Even before mergers, airlines such as Lufthansa, Delta Air Lines, United Airlines and British Airways were exhibiting DRS, which suggest that some of these airlines aim to increase their market power over operating at the optimal scale of operations.

5.6 Directional distance function—Chambers et al. (1996)

The DDF introduced by Chambers et al. (1996, 1998) is another important tool in production theory (Färe and Grosskopf, 2000). In essence, the DDF yields Shephard-type input and output distance functions, although Farrell technical efficiency can also be derived therefrom. Let us denote technology T using input vectors $x \in \Re_+^N$ to produce output vectors $y \in \Re_+^M$,

$$T = \{(x,y): \ x \text{ can produce } y\}. \tag{5.15}$$

Following Chambers et al. (1998) and denoting $g = (-g_x, g_y)$ as a directional vector, the *DDF* for technology T is defined as

$$\overrightarrow{D}_T(x, y; \ -g_x, \ g_y) = \sup\{\beta: \ (x - \beta g_x, \ y + \beta g_y) \in T\} \tag{5.16}$$

The *DDF* is a general functional representation of the technology and expressed as

$$\overrightarrow{D}_T(x, y; \ -g_x, \ g_y) \geq 0 \text{ if an only if}(x, y) \in T. \tag{5.17}$$

If the direction vector is specified as $g = (-g_x, g_y) = (x, 0)$, then the Shephard's input distance function is

$$D_i(x, y) = \max\{D > 0 | (x/D, y) \in T\} = 1/E(x, y) \tag{5.18}$$

where $E(x,y) = \min\{E > 0 | \ (Ex, y) \ \varepsilon \ T\}$ is the input-oriented Farrell efficiency based on technology T where inputs x produce outputs y. Hence, the Shephard input distance function is the inverse of the Farrell input-oriented efficiency.

If the direction vector is specified as $g = (-g_x, g_y) = (0, y)$, then the Shephard's output distance function is

$$D_o(x, y) = \min\{D > 0 | (x/D, y) \in T\} = 1/F(x, y) \tag{5.19}$$

where $F(x, y) = \max\{F > 0 | \ (x, Fy) \ \varepsilon \ T\}$ is the output-oriented Farrell efficiency based on technology T where inputs x produce outputs y. We note that the Shephard output distance function is the inverse of the Farrell output-oriented efficiency.

Given the aforementioned functions, Färe and Grosskopf (2000) showed the relationships between *DDF* and the input and output distance functions as Eqs. (5.20) and (5.21), respectively:

$$\overrightarrow{D}_T(x, y; \ x, 0) = 1 - 1/D_i(y, x) \tag{5.20}$$

$$\overrightarrow{D}_T(x, y; \ 0, y) = (1/D_o(x, y)) - 1 \tag{5.21}$$

TABLE 5.19 R script for Chambers et al. (1998) DDF.

```
## Input-oriented Directional distance function (CRS) – Chamber, Chung and Fare (1996)
library(Benchmarking)
## Load data (Reading from C drive and folder of data file)
data <- read.table(file="C:/Airline/Delivery_DDF.txt", header=T,sep="\t")
attach(data)
X = cbind(x1,x2,x3)
Y = cbind(y1,y2)
z <- c(1,1,1) #This refers to the number of inputs (direction) under input-orientation
e <- dea.direct(X,Y,DIRECT=z, RTS="crs", ORIENTATION = "in")   ⌉⌐ RTS and Orientation options, see 'Bogetoft
data.frame(dmu,Farrell_Di=e$eff, objval=e$objval)              ⌊  and Otto (2015).

## Output-oriented Directional distance function (CRS) – Chamber, Chung and Fare (1996)
library(Benchmarking)
## Load data (Reading from C drive and folder of data file)
data <- read.table(file="C:/Airline/Delivery_DDF.txt", header=T,sep="\t")
attach(data)
X = cbind(x1,x2,x3)
Y = cbind(y1,y2)
z <- c(1,1) #This refers to the number of outputs (direction) under output-orientation
e <- dea.direct(X,Y,DIRECT=z, RTS="crs", ORIENTATION = "out")
data.frame(dmu,Farrell_Do=e$eff, objval=e$objval)
```

5.6.1 R script for Chambers et al. (1998) directional distance function

The R script to generate Chambers et al. (1998) is based on the 'Benchmarking' package developed by Bogetoft and Otto (2011, 2015). The script shown in Table 5.19 presents the input- and output-oriented distance functions.

5.6.2 Interpretation of directional distance function results

The results shown in Table 5.20 are based on the 'delivery' data of Appendix C Table C.2 using the input-oriented DDF CRS R script shown in Table 5.19.

To obtain the overall Farrell input efficiency scores, we take the average of the Farrell_di.x1 + Farrell_di.x2 + Farrell_di.x3. As for the overall Farrell output efficiency scores, although we do not show the results, simply take the average of the Farrell_di.y1 + Farrell_di.y1 + Farrell_di.y2 when using the output-oriented DDF CRS R script shown in Table 5.19. The Farrell input efficiency scores are presented in Table 5.21.

The results from Table 5.21 show the same airlines that are efficient under standard DEA. As for the inefficient airlines, the rankings have somewhat changed. For example, when compared with the standard DEA CRS efficiency scores of Table 3.10 or Table 5.9, the worst performing airline is now TAM instead of ANA. The reason for this is because of the direction vector. If the direction vector was $g_x = x0$, and $g_y = y0$, then the directional distance function model will be equivalent to the radial model. Hence, the choice of the direction plays a significant role in determining the efficiency of airlines and should thus be considered only if the directional distance approach is

TABLE 5.20 Results for input distance function (CRS).

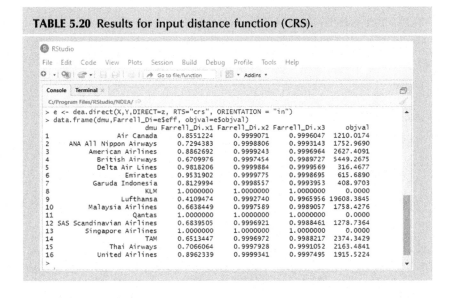

```
> e <- dea.direct(X,Y,DIRECT=z, RTS="crs", ORIENTATION = "in")
> data.frame(dmu,Farrell_Di=e$eff, objval=e$objval)
                      dmu Farrell_Di.x1 Farrell_Di.x2 Farrell_Di.x3      objval
1              Air Canada     0.8551224     0.9999071     0.9996047   1210.0174
2     ANA All Nippon Airways 0.7294383     0.9998806     0.9993143   1752.9690
3        American Airlines   0.8862692     0.9999243     0.9996964   2627.4091
4         British Airways    0.6709976     0.9997454     0.9989727   5449.2675
5         Delta Air Lines    0.9818206     0.9999884     0.9999569    316.4677
6                Emirates    0.9531902     0.9999775     0.9998695    615.6890
7        Garuda Indonesia    0.8129994     0.9998557     0.9993953    408.9703
8                    KLM    1.0000000     1.0000000     1.0000000      0.0000
9              Lufthansa    0.4109474     0.9992740     0.9965956  19608.3845
10      Malaysia Airlines    0.6638449     0.9997589     0.9989057   1758.4276
11                 Qantas    1.0000000     1.0000000     1.0000000      0.0000
12 SAS Scandinavian Airlines 0.6839505     0.9996921     0.9988461   1278.7364
13      Singapore Airlines   1.0000000     1.0000000     1.0000000      0.0000
14                    TAM    0.6513447     0.9996972     0.9988217   2374.3429
15            Thai Airways    0.7066064     0.9997928     0.9991052   2163.4841
16         United Airlines    0.8962339     0.9999341     0.9997495   1915.5224
>
```

TABLE 5.21 Farrell input efficiency scores (CRS).

	Input distance function	Rank	Standard DEA (Table 3.10)	Rank
Air Canada	0.9515	8	0.9098	12
ANA All Nippon Airways	0.9095	10	0.7294	16
American Airlines	0.9620	7	0.9282	11
British Airways	0.8899	13	0.9097	13
Delta Air Lines	0.9939	4	0.9887	4
Emirates	0.9843	5	0.9779	5
Garuda Indonesia	0.9374	9	0.9333	10
KLM	1.0000	1	1.0000	1
Lufthansa	0.8023	16	0.9448	8
Malaysia Airlines	0.8875	14	0.9765	6

Continued

TABLE 5.21 Farrell input efficiency scores (CRS).—cont'd

	Input distance function	Rank	Standard DEA (Table 3.10)	Rank
Qantas	1.0000	1	1.0000	1
SAS Scandinavian Airlines	0.8942	12	0.9691	7
Singapore Airlines	1.0000	1	1.0000	1
TAM	0.8833	15	0.8767	15
Thai Airways	0.9018	11	0.8819	14
United Airlines	0.9653	6	0.9356	9

meaningful in the context of measuring airline performance. In many studies, DDF has become a widely used method in measuring environmental performance incorporating bad outputs. We discuss more on this in Chapter 6.

5.7 Conclusion

In this chapter, we cover some extensions of DEA models, namely, the MF DEA, SBM, superefficiency DEA, potential gains DEA, and DDF. The purpose of introducing variations of DEA models is to provide the reader options to consider when results may be questionable or model choice to suit the purpose of their study. For example, if results exhibit considerable efficient DMUs derived from limited data, then one may consider alternative DEA models to provide some form of discrimination, such as the SBM of Tone (2001) or superefficiency DEA models. At times, comparisons between airlines could be motivated by the need to compare business models of airlines (i.e. mainstream airlines vs low-cost airlines; or between regions), which would require the MF DEA. This form of analysis provides airline management a way to learn from other airlines success and what may work for them. At times, mergers between airlines could be beneficial, but because of the cost involved in the initial stage, potential gains DEA would provide stakeholders a quantitative outcome from a merger. Finally, we introduce the DDF to illustrate an alternate efficiency measurement to the standard DEA. It is also mentioned as a prelude to the Malmquist–Luenberger productivity index, which is discussed in Chapter 6.

Appendix C

Tables C.1 and C.2.

TABLE C.1 Data set for 'Delivery_Meta'.

DMU	Flight personnel x1	Available tonne kilometres (thousands) x2	Fuel burn (tonnes) x3	Passenger tonne kilometres performed (thousands) y1	Freight and mail tonne kilometres performed (thousands) y2
Air Canada	8352	13,028,613	3,060,770.4	6,420,786	1,157,081
American Airlines	23,102	34,707,729	8,654,892.9	17,866,791	2,417,898
Delta Air Lines	17,408	27,292,425	7,349,946.5	14,571,329	1,671,083
TAM	6810	7,840,248	2,015,096.4	3,935,997	155,797
United Airlines	18,460	29,065,589	7,647,835.3	14,645,900	2,340,509
ANA All Nippon Airways	6479	14,683,532	2,556,513.8	4,286,268	2,059,289
Emirates	13,153	27,369,447	4,717,271.6	11,276,662	6,531,110
Garuda Indonesia	2187	2,834,184	676,346.5	1,514,745	282,129
Malaysia Airlines	5231	7,292,543	1,606,904.0	2,997,171	2,072,022
Qantas	12,156	17,368,244	3,156,052.3	9,945,797	2,623,457
Singapore Airlines	9467	21,286,125	3,513,669.0	7,733,939	6,559,460
Thai Airways	7374	10,441,041	2,417,856.2	4,725,671	2,157,255
British Airways	16,563	21,401,581	5,304,411.5	10,079,586	4,438,214
KLM	8101	15,090,771	3,027,818.2	7,347,192	4,093,466
Lufthansa	33,288	27,007,957	5,759,785.6	12,398,774	6,928,900
SAS Scandinavian Airlines	4046	4,152,670	1,108,178.3	2,304,528	344,994

TABLE C.2 Dataset for 'Delivery_DDF'.

DMU	Flight personnel $x1$	Available tonne kilometres (thousands) $x2$	Fuel burn (tonnes) $x3$	Passenger tonne kilometres performed (thousands) $y1$	Freight and mail tonne kilometres performed (thousands) $y2$
Air Canada	8352	13,028,613,000	3,060,770	6,420,786,000	1,157,081,000
ANA All Nippon Airways	6479	14,683,532,000	2,556,514	4,286,268,000	2,059,289,000
American Airlines	23,102	34,707,729,000	8,654,893	17,866,791,000	2,417,898,000
British Airways	16,563	21,401,581,000	5,304,411	10,079,586,000	4,438,214,000
Delta Air Lines	17,408	27,292,425,000	7,349,946	14,571,329,000	1,671,083,000
Emirates	13,153	27,369,447,000	4,717,272	11,276,662,000	6,531,110,000
Garuda Indonesia	2187	2,834,184,000	676,346.5	1,514,745,000	282,129,000
KLM	8101	15,090,771,000	3,027,818	7,347,192,000	4,093,466,000
Lufthansa	33,288	27,007,957,000	5,759,786	12,398,774,000	6,928,900,000
Malaysia Airlines	5231	7,292,543,000	1,606,904	2,997,171,000	2,072,022,000
Qantas	12,156	17,368,244,000	3,156,052	9,945,797,000	2,623,457,000
SAS Scandinavian Airlines	4046	4,152,670,000	1,108,178	2,304,528,000	344,994,000
Singapore Airlines	9467	21,286,125,000	3,513,669	7,733,939,000	6,559,460,000
TAM	6810	7,840,248,000	2,015,096	3,935,997,000	155,797,000
Thai Airways	7374	10,441,041,000	2,417,856	4,725,671,000	2,157,255,000
United Airlines	18,460	29,065,589,000	7,647,835	14,645,900,000	2,340,509,000

References

Andersen, P., Petersen, N.C., 1993. A procedure for ranking efficient units in data envelopment analysis. Management Science 39 (10), 1261–1264.

Battese, G.E., Rao, D.S.P., 2002. Technology potential, efficiency and a stochastic metafrontier function. International Journal of Business and Economics 1 (2), 1–7.

Battese, G.E., Rao, D.S.P., O'Donnell, C.J., 2004. A metafrontier production function for estimation of technical efficiencies and technology potentials for firms operating under different technologies. Journal of Productivity Analysis 21, 91–103.

Bogetoft, P., Otto, L., 2011. Benchmarking with DEA, SFA, and R. Springer, New York.

Bogetoft, P., Otto, L., 2015. Benchmark and Frontier Analysis Using DEA and SFA. Package 'Benchmarking'. Version 0.26. https://cran.rproject.org/web/packages/Benchmarking/Benchmarking.pdf.

Bogetoft, Wang, 2005. Estimating the potential gains from mergers. Journal of Productivity Analysis 23, 145–171.

Chambers, R.G., Chung, Y., Färe, R., 1996. Benefit and distance functions. Journal of Economic Theory 70, 407–419.

Chambers, R.G., Chung, Y., Färe, R., 1998. Profit, directional distance function, and Nerlovian efficiency. Journal of Optimization Theory and Applications 98 (2), 351–364.

Charnes, A., Cooper, W.W., 1962. Programming with linear fractional functional. Naval Research Logistics Quarterly 15, 333–334.

Chen, Y., 2004. Ranking efficient units in DEA. Omega 32, 213–219.

Chen, Y., 2005. Measuring super-efficiency in DEA in the presence of infeasibility. European Journal of Operational Research 161, 545–551.

Chen, Y., Du, J., Huo, J.Z., 2013a. Super-efficiency based on a modified directional distance function. Omega 41, 621–625.

Chen, Y., Djamasbi, S., Du, J., Lim, S., 2013b. Integer-valued DEA super-efficiency based on directional distance function with an application of evaluating mood and its impact on performance. International Journal of Production Economics 46 (2), 550–556.

Chen, Y., Liang, L., 2011. Super-efficiency DEA in the presence of infeasibility: one model approach. European Journal of Operational Research 213, 359–360.

Coll-Serrano, V., Bolós, V., Benítez, R., 2018. Conventional and Fuzzy Data Envelopment Analysis. R package 'deaR'. Version 1.3. https://cran.r-project.org/web/packages/deaR/deaR.pdf.

Cook, W.D., Liang, L., Zha, Y., Zhu, J., 2009. A modified super-efficiency DEA model for infeasibility. Journal of the Operational Research Society 60, 276–281.

Cooper, W.W., Seiford, L.M., Tone, K., 2007. Data Envelopment Analysis A Comprehensive Text with Models, Applications, References and DEA-Solver Software, Second edition. Springer, New York.

Du, J., Chen, C.-M., Chen, Y., Cook, W.D., Zhu, J., 2012. Additive super-efficiency in integer-valued data envelopment analysis. European Journal of Operational Research 218 (1), 186–192.

Du, J., Liang, L., Zhu, J., 2010. A slacks-based measure of super-efficiency in data envelopment analysis: a comment. European Journal of Operational Research 204, 694–697.

Färe, R., Grosskopf, S., 2000. Theory and application of directional distance functions. Journal of Productivity Analysis 13, 93–103.

Goergen, M., Renneboog, L., 2004. Shareholder wealth effects of European domestic and cross-border takeover bids. European Financial Management 10 (1), 9–45.

Lee, H.-S., Chu, C.W., Zhu, J., 2011. Super-efficiency DEA in the presence of infeasibility. European Journal of Operational Research 212, 141−147.

Lee, H.-S., Zhu, J., 2012. Super-efficiency infeasibility and zero data in DEA. European Journal of Operational Research 216, 429−433.

Lovell, C.A.K., Rouse, A.P.B., 2003. Equivalent standard DEA models to provide super-efficiency scores. Journal of the Operational Research Society 54, 101−108.

Neary, P., 2007. Cross-Border mergers as instruments of comparative advantage. The Review of Economic Studies 74, 1229−1257.

O'Donnell, C.J., 2018. Productivity and Efficiency Analysis: An Economic Approach to Measuring and Explaining Managerial Performance. Springer, Singapore.

O'Donnell, C.J., Rao, D.S.P., Battese, G.E., 2008. Metafrontier frameworks for the study of firm-level efficiencies and technology ratios. Empirical Economics 34, 231−255.

O'Donnell, C.J., 2010. Measuring and decomposing agricultural productivity and profitability change. The Australian Journal of Agricultural and Resource Economics 54 (4), 527−560.

Prince, J.Y., Simon, D.H., 2017. The impact of mergers on quality provision: evidence from the airline industry. The Journal of Industrial Economics 65, 336−362.

Ray, S.C., 2008. The directional distance function and measurement of super-efficiency: an application to airlines data. Journal of the Operational Research Society 59 (6), 788−797.

Seiford, L.M., Zhu, J., 1999. Infeasibility of super-efficiency data envelopment analysis models. INFOR: Information Systems and Operational Research 37 (2), 174−187.

Shott, T., Lim, D.J., 2015. TFDEA (Technology Forecasting using DEA). Package 'TFDEA'. Version 0.9.8. https://mran.microsoft.com/snapshot/2014-11-17/web/packages/TFDEA/index.html.

Tone, K., 2001. A slacks-based measure of efficiency in data envelopment analysis. European Journal of Operational Research 130, 498−509.

Tone, K., 2002. A slacks-based measure of super-efficiency in data envelopment analysis. European Journal of Operational Research 143, 32−41.

Chapter 6

Measuring airline performance: incorporating bad outputs

6.1 Introduction

In Chapters 3 and 4, we measured airline efficiency and productivity, respectively. We considered three inputs and two outputs for the *delivery* model, as this production function is key to airline performance and success. However, as mentioned in the seminal work of Koopmans (1951), a production process may also generate bad outputs such as smoke pollution or waste. This production process reflects the nature of aviation as not only airlines convey passengers and freight and mail, but also the consumption of fuel inevitably generates greenhouse gases (e.g. carbon dioxide [CO_2] emissions). This statement suggests that airlines not only produce passenger tonne km performed (PTKP) and freight and mail tonne km performed (FTKP) but also produce CO_2 emissions. In the literature on airline performance, PTKP and FTKP are considered good outputs, whereas CO_2 emission is a bad output. The literature also recognizes that good and bad outputs are jointly produced and that it can be costly to reduce bad output (i.e. weak disposability) because of resource and technology constraint (Färe et al., 1989, 2001).

While there have been advancements in new and improved fuel-efficient jet engines in recent years, the rate of technological development may struggle to keep pace with rising global demand even in post-COVID-19 era. The International Civil Aviation Organization (ICAO) forecasted for 2018–2050 that the annual growth for passenger traffic (measured by revenue passenger kilometres) in post-COVID-19 was 3.6 percent, downgraded from 4.2 percent (https://www.icao.int/sustainability/Pages/Post-Covid-Forecasts-Scenarios.aspx). Even with the downgraded forecast, the increase in passenger traffic will increase the amount of CO_2 emissions.

Incorporating bad outputs into the airline production framework is now more relevant especially after the 39th ICAO Assembly agreement on the

Productivity and Efficiency Measurement of Airlines. https://doi.org/10.1016/B978-0-12-812696-7.00001-5

155

adoption of CORSIA (Carbon Offsetting and Reduction Scheme for International Aviation), a global offsetting scheme, to address CO_2 emissions from international aviation.[1] To date, there are only a handful of studies that incorporate CO_2 emissions as bad outputs in airline efficiency and productivity analysis. These include Lee et al. (2015), Cui et al. (2016), Lee et al. (2017) and Liu et al. (2017). This chapter thus aims to evaluate airline performance incorporating both good and bad outputs.

The data set used in this chapter is similar to the *delivery* model in Chapter 3 except this time we include a bad output, CO_2 emissions, and we aggregate the good outputs, PTKP and FTKP, into one good output (TKP). We do this mainly because we intend to keep the number of outputs as 2 to the rule of thumb regarding the minimum number of DMUs and x's and y's required for analysis to provide sufficient discrimination (see Table 3.1). Furthermore, because the main focus is to reduce CO_2 emissions, this chapter adopts an output orientation model as it is more meaningful for airlines to reduce the bad output with given inputs. The data set is shown in Appendix D.1. Data for CO_2 emissions are obtained from RDC Aviation (http://www.rdcaviation.com/). RDC Aviation data are based on the International Air Transport Association (IATA) Scheduled Reference Service (SRS) database, which contains over 99% of all flight schedules worldwide, thus ensuring that the data reflect those filed by the airlines themselves and align with the IATA World Air Transport Statistics (WATS) database.

The chapter is divided into five sections. Following the introduction in Section 6.1, we look at the various arguments on how environmental DEA technology is modelled in Section 6.2. Section 6.3 describes Seiford and Zhu (2002) transformation approach. Section 6.4 describes Zhou et al. (2008) Environmental DEA model. Section 6.5 describes Tone's SBM with bad outputs model in Cooper et al. (2007). Section 6.6 describes Chung et al. (1997) Malmquist–Luenberger (ML) model. This chapter concludes with some brief comments in Section 6.7.

6.2 Environmental DEA technology model

The treatment of undesirable outputs (sometimes referred to as bad outputs) in DEA has received extensive attention as evident in the extant literature detailed in Halkos and Petrou (2019). The incorporation of bad outputs into performance measurement can be categorized into three groups. The first group that includes studies such as Berg et al. (1992), Reinhard et al. (1999), Hailu and Veeman (2001) and Korhonen and Luptacik (2004) treats bad outputs as inputs in the production process. Seiford and Zhu (2002), however, argue that their approach distorts the production process and thus fails to

---

1. CORSIA came into effect on 1 January 2019.

reflect the real production process. The second group that includes studies such as Ali and Seiford (1990), Scheel (2001), Seiford and Zhu (2002) and Hua et al. (2007) is based on the translation invariance property of DEA via data transformation of undesirable outputs into positive values and then evaluating the environmental efficiency. The limitation of this approach is that due to strong convexity constraints, it can only be solved under variable returns to scale. The third group considers the weak disposable reference technology introduced by Färe et al. (1989) and is based on a piece-wide linear production technology characterized within an environmental DEA technology. Their study treats good and bad outputs asymmetrically, resulting in an enhanced hyperbolic output efficiency measure, and is solved via nonlinear programming. Hailu and Veeman (2001) argued against Färe et al. (1989) approach that assuming weak disposability leaves the impact of undesirable outputs on efficiency undetermined. Fare and Grosskopf (2004) disagreed with Hailu and Veeman (2001) as weak disposability DEA model is consistent with physical laws and allows the treatment of bad outputs showing the opportunity cost of reducing them. This strand of treatment of bad outputs has since gained recognition in recent times with new developments in studies such as Färe et al. (1993), Färe and Grosskopf (2004), Yaisawarng and Klein (1994), Chung et al. (1997), Zofio and Prieto (2001), Zhou et al. (2008) and Tone and Tsutsui (2011). From the extant literature of environmental DEA models, we consider four approaches—Seiford and Zhu (2002) transformation approach, Zhou et al. (2008) radial efficiency measure, Tone's SBM approach in Cooper et al. (2007) and Chung et al. (1997) Malmquist—Luenberger productivity index. The first three approaches are cross-sectional analysis, whereas the fourth approach is a panel-data analysis.

6.3 Seiford and Zhu (2002) transformation approach

Seiford and Zhu's (2002) treatment of good and bad outputs concurrently follows the linear monotone decreasing transformation based upon the classification invariance concept in Ali and Seiford (1990) and preserves the linearity and convexity of the DEA model. Starting with the following DEA data domain:

$$\begin{bmatrix} Y \\ -X \end{bmatrix} = \begin{bmatrix} Y^g \\ Y^b \\ -X \end{bmatrix} \tag{6.1}$$

where X represents the inputs and Y^g and Y^b represent the corresponding good and bad outputs, respectively. To increase Y^g while reducing Y^b,

Seiford and Zhu (2002) multiplied each bad output by negative one and then find and apply a proper translation vector value w to convert all negative bad outputs into positives $\left(Y^{-b} = -Y_j^b + w > 0\right)$, which results in the following domain:

$$\begin{bmatrix} Y \\ -X \end{bmatrix} = \begin{bmatrix} Y^g \\ Y^{-b} \\ -X \end{bmatrix} \tag{6.2}$$

Based on Eq. (6.2), Seiford and Zhu (2002) modified Banker et al. (1984) input-oriented BCC model to the following linear program:

$$\max \theta$$

$$\text{subject to } \sum_{j=1}^{n} z_j y_j^g \geq \theta y_o^g$$

$$\sum_{j=1}^{n} z_j y_j^{-b} \geq \theta y_o^{-b}$$

$$\sum_{j=1}^{n} z_j x_j \leq x_o$$

$$\sum_{j=1}^{n} z_j = 1$$

$$z_j \geq 0, \quad j = 1, 2, \ldots, n \tag{6.3}$$

Here, θ is the efficiency score of the DMU, y_j^g and y_j^b are the good and bad outputs of airline j, respectively. x_j is the j-th input and z_j is the weight of airline j, and x_o and y_o represent the input and output vectors for all airlines.

6.3.1 R script for Seiford and Zhu (2002) model

The R script for Seiford and Zhu (2002) model shown in Table 6.1 is based on the R package 'deaR' developed by Coll-Serrano et al. (2018).

The data used in the model are drawn from Appendix D.1, which consist of three inputs, one good output and one bad output. The translation parameter $= 25,000,000,000$ to convert all bad outputs (in negative terms) into positive.

TABLE 6.1 R script for Seiford and Zhu (2002) model.

```
## load deaR
library(deaR)

## Load data
data <- read.table(file="C:/Airline/Delivery_bad.txt", header=T,sep="\t")
attach(data)
data_goodandbad <- read_data(data,ni = 3,no = 2,ud_outputs=2)
data_good <- read_data(data,ni = 3,no = 1)

result1 <- model_basic(data_goodandbad,orientation="oo",rts="vrs",vtrans_
o=27400000)
result2 <- model_basic(data_good,orientation="oo",rts="vrs")

Seiford_Zhu <- efficiencies(result1)
Standard_DEA<- efficiencies(result2)
data.frame(Seiford_Zhu, Standard_DEA)
```

> "data,ni = 3,no = 2,ud_outputs=2" is read as follows:
>
> The first three (3) columns are inputs, followed by two (2) columns of outputs, of which the second output (2) is the bad output.

6.3.2 Interpretation of Seiford and Zhu (2002) results

Table 6.2 presents the efficiency estimates using Seiford and Zhu (2002) model and standard DEA technical efficiency estimates under VRS output orientation. It is not possible to compare the Seiford_Zhu results with Table 3.10 results because the latter uses two good outputs (i.e. $y1$ and $y2$), whereas the current model takes the aggregation of $y1$ and $y2$ into one good output.

The results from Table 6.2 show that both DEA models exhibit the same seven efficient airlines. The results support the finding in Färe et al. (1989) that DEA models that fail to credit airlines for reducing bad output (i.e. reducing carbon emissions) can severely distort an airline's eco-efficiency performance. Seiford and Zhu (2002) model is, however, limited to the VRS model due to strong convexity constraints. It should also be mentioned that Seiford and Zhu (2002) model is dependent on the size of the translation vector w in $\left(Y^{-b} = -Y_j^b + w \right)$ Different values of w were used and resulted in variation in the efficiency estimates, except for Garuda Indonesia, which maintained an efficiency estimate of 1. We note that as w gets larger, the number of efficient airlines increases.

6.4 Zhou et al. (2008) environmental DEA model

Zhou et al. (2008) developed a radial DEA-based model for measuring environmental performance under different technology sets, incorporating both

TABLE 6.2 DEA technical efficiency and efficiency estimates using Seiford and Zhu (2002) model, VRS output orientation.

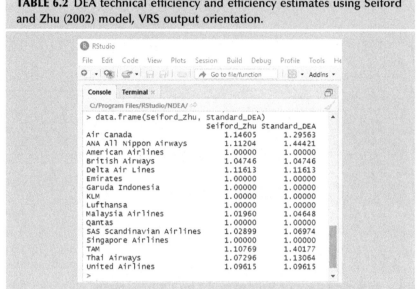

good and bad outputs under weak disposability assumption. Referred to as environmental performance index (EPI) models, these models based on undesirable output orientation provide DEA technology under CRS, NIRS, VRS and MEI (mixed environmental performance index) and are described in the following. The production technology set $P(x)$ can then be characterized by the output set:

$$P(x) = \{(y, b); x \text{ can produce } (y, b)\}. \tag{6.4}$$

According to Chung et al. (1997) and Färe and Grosskopf (2004), $P(x)$ is defined as an 'environmental output set' if $P(x)$ is bounded and closed set and satisfies the following two properties:

Property 1: Outputs y and b are weakly disposable. If $(y, b) \in P(x)$ and $0 \le \theta \le 1$, then $(\theta y\ \theta b) \in P(x)$.

Property 2: Good outputs and bad outputs are null-joint. If $(y, b) \in P(x)$ and $b = 0$, then $y = 0$.

6.4.1 Pure environmental performance index (EPI$_{CRS}$)

Suppose there are n DMUs ($n = 1, 2, ..., N$). For each DMU$_n$, the observed data on the vectors of inputs, good outputs and bad outputs are

$x_n = (x_{1n}, x_{2n}, ..., x_{Mn})$, $y_n = (y_{1n}, y_{2n}, ..., y_{Un})$ and $b_n = (b_{1n}, b_{2n}, ..., b_{Kn})$, respectively. Then, the EPI_{CRS} can be expressed as follows:

$$EPI_{CRS} = \min \lambda$$

$$\text{subject to} \quad \sum_{n=1}^{N} z_n x_{mn} \leq x_{m0}, \quad m = 1, 2, ..., M$$

$$\sum_{n=1}^{N} z_n y_{un} \geq y_{u0}, \quad u = 1, 2, ..., U$$

$$\sum_{n=1}^{N} z_n b_{kn} = \lambda b_{k0}, \quad k = 1, 2, ..., K$$

$$z_n \geq 0, \quad n = 1, 2, ..., N \tag{6.5}$$

where z_n is the intensity variable. Eq. (6.5) is considered 'pure' because only adjustments to bad outputs, and none to good outputs, are allowed.

6.4.2 NIRS environmental performance index (EPI_{NIRS})

$$EPI_{NIRS} = \min \lambda$$

$$\text{subject to} \quad \sum_{n=1}^{N} z_n x_{mn} \leq x_{m0}, \quad m = 1, 2, ..., M$$

$$\sum_{n=1}^{N} z_n y_{un} \geq y_{u0}, \quad u = 1, 2, ..., U$$

$$\sum_{n=1}^{N} z_n b_{kn} = \lambda b_{k0}, \quad k = 1, 2, ..., K$$

$$\sum_{n=1}^{N} z_n \leq 1$$

$$z_n \geq 0, \quad n = 1, 2, ..., N \tag{6.6}$$

6.4.3 VRS environmental performance index (EPI$_{VRS}$)

$$EPI_{VRS} = \min \lambda$$

$$\text{subject to} \quad \sum_{n=1}^{N} z_n x_{mn} \leq \beta x_{m0}, \quad m = 1, 2, \ldots, M$$

$$\sum_{n=1}^{N} z_n y_{un} \geq y_{u0}, \quad u = 1, 2, \ldots, U$$

$$\sum_{n=1}^{N} z_n b_{kn} = \lambda b_{k0}, \quad k = 1, 2, \ldots, K$$

$$\sum_{n=1}^{N} z_n = \beta$$

$$\beta \leq 1, \; z_n \geq 0, \quad n = 1, 2, \ldots, N \tag{6.7}$$

6.4.4 Mixed environmental performance index

$$MEI_{VRS} = \min \lambda$$

$$\text{subject to} \quad \sum_{n=1}^{N} z_n x_{mn} \leq \beta x_{m0}, \quad m = 1, 2, \ldots, M$$

$$\sum_{n=1}^{N} z_n y_{un} \geq y_{u0}, \quad u = 1, 2, \ldots, U$$

$$\sum_{n=1}^{N} z_n b_{kn} = \lambda b_{k0}, \quad k = 1, 2, \ldots, K$$

$$\sum_{n=1}^{N} z_n = \beta$$

$$z_n \geq 0, \quad n = 1, 2, \ldots, N \tag{6.8}$$

Eq. (6.8) is also referred to as the aggregated and standardized EPI.

EPI and MEI lie in the interval $(0, 1)$.[2] An airline is eco-efficient if $EPI = 1$ and eco-inefficient if $EPI < 1$. Hence, the closer an airline's EPI is to 1, the better its environmental performance indicating its efficiency in controlling pollutants. MEI could also be used to measure the efficiency of an airline scale of production with regards to good outputs and bad outputs. If $MEI = 1$, an airline is deemed to be at its ideal environmental scale size. If $MEI < 1$, it indicates that an airline is not operating at its ideal environmental scale size.

6.4.5 R script for Zhou et al. (2008) environmental DEA model

The R script Zhou et al. (2008) environmental DEA model is shown in Table 6.3 and uses the data set shown in Appendix D.1.[3]

6.4.6 Discussion of results

In this section, we describe and interpret the results of our environmental DEA efficiency estimates. EPI of airlines is presented in Table 6.4 for each of the four DEA technology sets.

When we compare the results of Table 6.4 with the reciprocal of the standard DEA efficiency scores of Appendix D.2, we observe that the EPI has lower efficiency scores than the standard DEA. This is a finding observed in Scheel (2001) that under standard DEA model, the efficiency score is over-estimated because one assumes that good outputs can be increased while ignoring bad outputs. The reality is that good and bad outputs are sometimes inseparable, as in the case of airlines. The results also reveal the same three airlines (KLM, Qantas and Singapore Airlines) maintaining their efficiency based on the bad output orientation model under CRS. This result supports Scheel (2001) nonseparating approach that some DMUs perform well with respect to either good or bad outputs, respectively. Under VRS, the results reveal more efficient airlines as expected because VRS measures pure technical efficiency and does not include scale efficiency. MEI under VRS reveals only Garuda Indonesia and Singapore Airlines as efficient indicating their ability to simultaneously adjust the good and bad outputs.

Table 6.4 also provides information regarding returns to scale. Each of the EPI can be used together to determine the returns to scale properties of a DMU. Following the idea of Färe et al. (1994), the reference DMU exhibits CRS if $EPI_{CRS} = EPI_{VRS} = 1$. If they are not, then one has to determine the

---

2. Note that output-orientated efficiency scores are bounded between 0 and 1 (Coelli et al., 1998). In many DEA studies that use output orientation, the efficiency estimates range from 1 to ∞. Nonetheless, we adopt the conventional approach and use the range between 0 and 1 mainly for comparisons.
3. Table 6.3 R script was written by Manh D. Pham (University of Queensland, Ph.D. candidate 2018).

TABLE 6.3 R script for Environmental DEA model Zhou et al. (2008).

```
## Load data
data <- read.table(file="C:/Airline/Delivery_bad.txt", header=T,sep="\t")
attach(data)
X <- cbind(x1,x2,x3)
Y <- cbind(y1)
B <- cbind(b1)

## WEAK DISPOSABILITY (WD)

wd.g = function(xo,yo,bo,Xref,Yref,Bref,rts){
 library(lpSolve)
 n = nrow(Xref)
 p = ncol(Xref)
 q = ncol(Yref)
 j = ncol(Bref)

if (rts=="crs") {
  obj.fun = c(1,rep(0,n))
  constr.x = cbind(matrix(0,p,1),t(Xref),-xo)
  constr.y = cbind(matrix(0,q,1),-t(Yref),yo)
  constr.lb = -diag(1+n)
  constr.b = cbind(matrix(0,j,1),t(Bref))
  constr = rbind(constr.x,constr.y,constr.lb,constr.b)
  constr.dir = c(rep("<=",p+q+1+n),rep("=",j))
  rhs = c(xo,rep(0,q+1+n),bo) }

else if (rts=="nirs") {
  obj.fun = c(1,rep(0,n))
  constr.x = cbind(matrix(0,p,1),t(Xref),-xo)
  constr.y = cbind(matrix(0,q,1),-t(Yref),yo)
  constr.lb = -diag(1+n)
  constr.lambda = cbind(0,matrix(1,1,n))
  constr.b = cbind(matrix(0,j,1),t(Bref))
  constr = rbind(constr.x,constr.y,constr.lb,constr.lambda,constr.b)
  constr.dir = c(rep("<=",p+q+1+n+1),rep("=",j))
  rhs = c(xo,rep(0,q+1+n),1,bo)  }

else {
  obj.fun = c(1,rep(0,n+1))
  constr.x = cbind(matrix(0,p,1),t(Xref),-xo)
  constr.y = cbind(matrix(0,q,1),-t(Yref),yo)
  constr.lb = -diag(2+n)
  constr.ub = cbind(matrix(0,1,1+n),1)
  constr.gamma = cbind(0,matrix(1,1,n),-1)
  constr.b = cbind(matrix(0,j,1),t(Bref),matrix(0,j,1))
  constr = rbind(constr.x,constr.y,constr.lb,constr.ub,constr.gamma,constr.b)
  constr.dir = c(rep("<=",p+q+2+n+1),rep("=",1+j))
  rhs = c(rep(0,p+q+2+n),1,0,bo) }

results = lp("max",obj.fun,constr,constr.dir,rhs)
wd.g = results$objval

if (rts=="vrs"){
 theta = results$solution[n+2]
 lambda = results$solution[2:(n+1)]/theta
 return(list(wd.g=wd.g,theta=theta,lambda=lambda)) }
```

```
else {
   lambda = results$solution[2:(n+1)]
   return(list(wd.g=wd.g,lambda=lambda))  }}

wd.g.all = function(X,Y,B,Xref,Yref,Bref,rts){
 m = nrow(X)
 n = nrow(Xref)
 if (rts=="vrs"){
  results = list(wd.g=rep(0,m),theta=rep(0,m),lambda=matrix(0,m,n))
  }
 else {
  results = list(wd.g=rep(0,m),lambda=matrix(0,m,n))
  }
 for (i in 1:m){
  temp = wd.g(X[i,],Y[i,],B[i,],Xref,Yref,Bref,rts)
  results$wd.g[i] = round(temp$wd.g,5)
  results$lambda[i,] = round(temp$lambda,5)
  if (rts=="vrs") { results$theta[i] = round(temp$theta,5) }
 }
 return(results)}

wd.i = function(xo,yo,bo,Xref,Yref,Bref,rts){
 library(lpSolve)
 n = nrow(Xref)
 p = ncol(Xref)
 q = ncol(Yref)
 j = ncol(Bref)

 if (rts=="crs"){
  obj.fun = c(1,rep(0,n))
  constr.x = cbind(-xo,t(Xref))
  constr.y = cbind(matrix(0,q,1),-t(Yref))
  constr.lb = -diag(1+n)
  constr.b = cbind(matrix(0,j,1),t(Bref))
  constr = rbind(constr.x,constr.y,constr.lb,constr.b)
  constr.dir = c(rep("<=",p+q+1+n),rep("=",j))
  rhs = c(rep(0,p),-yo,rep(0,1+n),bo) }

 else if (rts=="nirs"){
  obj.fun = c(1,rep(0,n))
  constr.x = cbind(-xo,t(Xref))
  constr.y = cbind(matrix(0,q,1),-t(Yref))
  constr.lb = -diag(1+n)
  constr.lambda = cbind(0,matrix(1,1,n))
  constr.b = cbind(matrix(0,j,1),t(Bref))
  constr = rbind(constr.x,constr.y,constr.lb,constr.lambda,constr.b)
  constr.dir = c(rep("<=",p+q+1+n+1),rep("=",j))
  rhs = c(rep(0,p),-yo,rep(0,1+n),1,bo)   }

 else {
  obj.fun = c(1,rep(0,n+1))
  constr.x = cbind(-xo,t(Xref),matrix(0,p,1))
  constr.y = cbind(matrix(0,q,1),-t(Yref),yo)
  constr.lb = -diag(2+n)
  constr.lambda = cbind(0,matrix(1,1,n),0)
  constr.b = cbind(matrix(0,j,1),t(Bref),-bo)
  constr = rbind(constr.x,constr.y,constr.lb,constr.lambda,constr.b)
  constr.dir = c(rep("<=",p+q+1+n+1),rep("=",1+j))
  rhs = c(rep(0,p+q+1+n),-1,1,rep(0,j))  }

 results = lp("min",obj.fun,constr,constr.dir,rhs)
 wd.i = results$objval
```

Continued

TABLE 6.3 R script for Environmental DEA model Zhou et al. (2008).—cont'd

```
lambda = results$solution[2:(n+1)]
if (rts=="vrs"){
  theta = 1/results$solution[2+n]
  return(list(wd.i=wd.i,theta=theta,lambda=lambda))
} else {
  return(list(wd.i=wd.i,lambda=lambda))   }}

wd.i.all = function(X,Y,B,Xref,Yref,Bref,rts){
  m = nrow(X)
  n = nrow(Xref)
  if (rts=="vrs"){
    results = list(wd.i=rep(0,m),theta=rep(0,m),lambda=matrix(0,m,n))
    for (i in 1:m) {
      temp = wd.i(X[i,],Y[i,],B[i,],Xref,Yref,Bref,rts)
      results$wd.i[i] = round(temp$wd.i,5)
      results$theta[i] = round(temp$theta,5)
      results$lambda[i,] = round(temp$lambda,5)
    }
  } else {
    results = list(wd.i=rep(0,m),lambda=matrix(0,m,n))
    for (i in 1:m) {
      temp = wd.i(X[i,],Y[i,],B[i,],Xref,Yref,Bref,rts)
      results$wd.i[i] = round(temp$wd.i,5)
      results$lambda[i,] = round(temp$lambda,5)
    }}
  return(results)}

wd.b = function(xo,yo,bo,Xref,Yref,Bref,rts){
  library(lpSolve)
  n = nrow(Xref)
  p = ncol(Xref)
  q = ncol(Yref)
  j = ncol(Bref)

  if (rts=="crs"){
    obj.fun = c(1,rep(0,n))
    constr.x = cbind(matrix(0,p,1),t(Xref))
    constr.y = cbind(matrix(0,q,1),-t(Yref))
    constr.b = cbind(bo,-t(Bref))
    constr.lb = -diag(1+n)
    constr = rbind(constr.x,constr.y,constr.b,constr.lb)
    constr.dir = c(rep("<=",p+q),rep("=",j),rep("<=",1+n))
    rhs = c(xo,-yo,rep(0,j),rep(0,1+n))  }

  else if (rts=="nirs"){
    obj.fun = c(1,rep(0,n))
    constr.x = cbind(matrix(0,p,1),t(Xref))
    constr.y = cbind(matrix(0,q,1),-t(Yref))
    constr.b = cbind(bo,-t(Bref))
    constr.lambda = cbind(0,matrix(1,1,n))
    constr.lb = -diag(1+n)
    constr = rbind(constr.x,constr.y,constr.b,constr.lambda,constr.lb)
    constr.dir = c(rep("<=",p+q),rep("=",j),rep("<",1+1+n))
    rhs = c(xo,-yo,rep(0,j),1,rep(0,1+n))  }

  else {
    obj.fun = c(1,rep(0,1+n))
    constr.x = cbind(matrix(0,p,1),t(Xref),-xo)
    constr.y = cbind(matrix(0,q,1),-t(Yref),matrix(0,q,1))
    constr.b = cbind(bo,-t(Bref),matrix(0,j,1))
```

TABLE 6.3 R script for Environmental DEA model Zhou et al. (2008).—cont'd

```
    constr.lambda = cbind(0,matrix(1,1,n),-1)
    constr.lb = -diag(1+n+1)
    constr.theta = cbind(matrix(0,1,1+n),1)
    constr = rbind(constr.x,constr.y,constr.b,constr.lambda,constr.lb,constr.theta)
    constr.dir = c(rep("<=",p+q),rep("=",j+1),rep("<=",1+n+1),"<=")
    rhs = c(rep(0,p),-yo,rep(0,j+1+1+n+1),1)  }

  results = lp("min",obj.fun,constr,constr.dir,rhs)
  wd.b = results$objval
  lambda = results$solution[2:(n+1)]
  if (rts=="vrs"){
    theta = 1/results$solution[2+n]
    return(list(wd.b=wd.b,theta=theta,lambda=lambda))
  } else {
    return(list(wd.b=wd.b,lambda=lambda))  }}


wd.b.all = function(X,Y,B,Xref,Yref,Bref,rts){
  m = nrow(X)
  n = nrow(Xref)
  if (rts=="vrs"){
    results = list(wd.b=rep(0,m),theta=rep(0,m),lambda=matrix(0,m,n))
    for (i in 1:m) {
      temp = wd.b(X[i,],Y[i,],B[i,],Xref,Yref,Bref,rts)
      results$wd.b[i] = round(temp$wd.b,5)
      results$theta[i] = round(temp$theta,5)
      results$lambda[i,] = round(temp$lambda,5)
    }
  } else {
    results = list(wd.b=rep(0,m),lambda=matrix(0,m,n))
    for (i in 1:m) {
      temp = wd.b(X[i,],Y[i,],B[i,],Xref,Yref,Bref,rts)
      results$wd.b[i] = round(temp$wd.b,5)
      results$lambda[i,] = round(temp$lambda,5)  } }
  return(results)}

wd.mei = function(xo,yo,bo,Xref,Yref,Bref){
  n = nrow(Xref)
  p = ncol(Xref)
  q = ncol(Yref)
  j = ncol(Bref)
  obj.fun = c(1,rep(0,1+n))
  constr.x = cbind(matrix(0,p,1),t(Xref),-xo)
  constr.y = cbind(matrix(q,1),-t(Yref),matrix(0,q,1))
  constr.b = cbind(bo,-t(Bref),matrix(0,j,1))
  constr.gamma = cbind(0,matrix(1,1,n),-1)
  constr.lb = -diag(1+n+1)
  constr = rbind(constr.x,constr.y,constr.b,constr.gamma,constr.lb)
  constr.dir = c(rep("<=",p+q),rep("=",j+1),rep("<=",1+n+1))
  rhs = c(rep(0,p),-yo,rep(0,j+1+1+n+1))
  results = lp("min",obj.fun,constr,constr.dir,rhs)
  return(results$objval)}

##MEI

wd.mei.all = function(X,Y,B,Xref,Yref,Bref){
  m = nrow(X)
  results = rep(0,m)
  for (i in 1:m){
    results[i] = wd.mei(X[i,],Y[i,],B[i,],Xref,Yref,Bref)  }
```

Continued

TABLE 6.3 R script for Environmental DEA model Zhou et al. (2008).—cont'd

```
return(results)}

## Measure EPI and MEI

EPIcrs = wd.b.all(X,Y,B,X,Y,B,"crs")
EPInirs = wd.b.all(X,Y,B,X,Y,B,"nirs")
EPIvrs = wd.b.all(X,Y,B,X,Y,B,"vrs")
MEI = wd.mei.all(X,Y,B,X,Y,B)

data.frame(dmu,EPIcrs=(EPIcrs$wd.b),EPInirs=(EPInirs$wd.b),EPIvrs=(EPIvrs$wd.b),MEI=(MEI))
```

TABLE 6.4 Environmental performance for four DEA technology models.

```
> data.frame(dmu,EPIcrs=round(EPIcrs$wd.b,5),EPInirs=round(EPInirs$wd.b,5),
EPIvrs=round(EPIvrs$wd.b,5),MEI=round(MEI,5))
                        dmu   EPIcrs  EPInirs   EPIvrs      MEI
1                 Air Canada  0.60861  0.60861  0.63849  0.63849
2     ANA All Nippon Airways  0.61016  0.61016  0.63625  0.63625
3          American Airlines  0.57615  1.00000  1.00000  0.57615
4            British Airways  0.67483  0.68681  0.68681  0.67280
5            Delta Air Lines  0.54324  0.56887  0.56887  0.54324
6                   Emirates  0.92799  1.00000  1.00000  0.92799
7           Garuda Indonesia  0.65309  0.65309  1.00000  1.00000
8                        KLM  1.00000  1.00000  1.00000  0.95833
9                  Lufthansa  0.84001  1.00000  1.00000  0.82489
10         Malaysia Airlines  0.78332  0.78332  0.88867  0.88867
11                    Qantas  1.00000  1.00000  1.00000  0.99606
12 SAS Scandinavian Airlines  0.58774  0.58774  0.77679  0.77679
13         Singapore Airlines  1.00000  1.00000  1.00000  1.00000
14                       TAM  0.49916  0.49916  0.56429  0.56429
15               Thai Airways  0.69979  0.69979  0.75588  0.75588
16           United Airlines  0.54599  0.58003  0.58003  0.54599
>
```

relationship between EPI_{NIRS} and EPI_{VRS}. If $EPI_{NIRS} < EPI_{VRS}$, the reference DMU, respectively, exhibits IRS. If $EPI_{NIRS} = EPI_{VRS}$, the reference DMU, respectively, exhibits DRS. From Table 6.4, IRS airlines include Air Canada, ANA, Garuda Indonesia, Malaysia Airlines, Scandinavian Airlines, TAM and

Thai Airways. DRS airlines include American Airlines, British Airways, Delta Air Lines, Emirates, Lufthansa and United Airlines. The results also seem to support the observation in Section 5.5.2 that airlines involved in mergers and acquisitions with the aim of expanding market power are also some of the worst performers in mitigating carbon emissions, except for Lufthansa.

6.5 Tone's SBM with bad outputs in Cooper et al. (2007)

A shortcoming of radial models is the neglect of nonradial input and output slacks (Tone, 2001). As the aforementioned models are radial models and deal with bad outputs, these models consequently suffer from slacks in bad outputs as they are not accounted for in the efficiency measure. To overcome the issue of slacks, one can incorporate bad output into the SBM model of Tone (2001) as attempted in Cooper et al. (2007), who extended on Tone's (2001) SBM model to include bad outputs. Extending on Eq. (5.6) in Chapter 5, the SBM efficiency, θ, under CRS is expressed as

$$\theta = \min_{\lambda, s^-, s^+} \frac{1 - \frac{1}{m}\sum_{i=1}^{m}\left(\frac{s_i^-}{x_{io}}\right)}{1 + \frac{1}{s_1 + s_2}\left[\sum_{r=1}^{s_1}\left(\frac{s_r^g}{y_{ro}^g}\right) + \sum_{r=1}^{s_2}\left(\frac{s_r^b}{y_{ro}^b}\right)\right]} \qquad (6.9)$$

subject to

$$x_{io} = X\lambda + s_i^- \quad (i = 1, \ldots, m)$$

$$y_{ro}^g = Y^g\lambda - s^g \quad (r = 1, \ldots, s_1)$$

$$y_{ro}^b = Y^b\lambda - s^b \quad (r = 1, \ldots, s_2)$$

$$\lambda \geq 0, s_i^- \geq 0 \ (\forall i), s^g \geq 0 \ (\forall r), s^b \geq 0 \ (\forall r)$$

where there are n DMUs, each using m inputs to produce good and bad outputs as represented by the vectors $x \in R^m$, $y^g \in R^{s_1}$ and $y^b \in R^{s_2}$, respectively. The matrices X, Y^g and Y^b are defined as $X = (x_1, x_2, \ldots, x_n) \in R^{m \times n}$, $Y^g = (y_1^g, y_2^g, \ldots, y_n^g) \in R^{s_1 \times n}$ and $Y^b = (y_1^b, y_2^b, \ldots, y_n^b) \in R^{s_2 \times n}$. The slacks $s^- = (s_1^-, s_2^-, \ldots, s_m^-) \in R^m$ and $s^b = (s_1^b, s_2^b, \ldots, s_n^b) \in R^{s_2}$ correspond to excesses in inputs and bad outputs, respectively, while $s^g = (s_1^g, s_2^g, \ldots, s_n^g) \in R^{s_1}$ reflects shortages in good outputs. $\lambda = (\lambda_1, \lambda_2, \ldots, \lambda_n)$ is the intensity vector. As Eq. (5.6) is in fractional form, Tone (2001) transforms it into a linear form using the Charnes–Cooper transformation (Charnes and Cooper 1962). For details of this, see Cooper et al. (2007).

TABLE 6.5 R script Tone's SBM with bad output model.

```
## load deaR
library(deaR)
## Load data
data <- read.table(file="C:/Airline/Delivery_bad.txt ", header=T,sep="\t")
attach(data)
data_example <- read_data(data,
ni = 3,
no = 2,
ud_outputs = 2)

result_crs <- model_sbmeff(data_example,orientation="no",rts = "crs")
result_vrs <- model_sbmeff(data_example,orientation="no",rts = "vrs")

eff_crs <- efficiencies(result_crs)
eff_vrs <- efficiencies(result_vrs)

data.frame(eff_crs, eff_vrs)
data.frame(targets(result_crs), slacks(result_crs))
data.frame(targets(result_vrs), slacks(result_vrs))
```

ni = 3, no = 2, ud_outputs=2 reads
as follows:
ni = 3: first three (3) columns are
inputs,
no = 2: next two (2) columns are
outputs,
ud_outputs = 2: the second output
(2) is the bad output.

Extra optional functions

6.5.1 R script for Tone's SBM with bad output

The R script for Tone's SBM with bad output model shown in Table 6.5 is based on the R package 'deaR' developed by Coll-Serrano et al. (2018).

The data used in the model are drawn from Appendix D.1, which consist of three inputs, one good output and one bad output.

The aforementioned instructions also allow for additional options that can be added to the lines 'Extra optional functions'. For example, weight_input = 1, weight_output = 1, orientation = c('no', 'io', 'oo') where 'no' refers to nonorientated, and rts = c('crs', 'vrs', 'nirs', 'ndrs', 'grs').

6.5.2 Interpretation of Tone's SBM with bad output results

Table 6.6 presents the efficiency estimates using Tone's SBM with bad output model under CRS and VRS nonorientation.

Table 6.6 results reveal KLM, Qantas and Singapore Airlines as efficient under CRS, which is the same as Zhou et al. (2008) EPI_{CRS} in Table 6.4. The same seven efficient airlines were also observed under VRS as those of Zhou et al. (2008) EPI_{VRS} in Table 6.4.

To further our interpretation of the results, we present the target and slacks input and output for CRS and VRS in Tables 6.7 and 6.8, respectively. Using Table 6.7 to illustrate, Lufthansa can raise its efficiency by reducing the number of flight personnel from by 20,009 to 13,278, reducing fuel burn by 812,134.2 tonnes to 4,947,651.4 tonnes and reducing CO_2 emissions by 2,563,907.7 tonnes to 15,619,735 tonnes.

Tone's SBM with bad output model described before assumes that the good and bad outputs are separable. However, it is sometimes observed that certain bad outputs are inseparable from the corresponding good outputs, suggesting

TABLE 6.6 Efficiency estimates using Tone's SBM with bad output model, CRS and VRS nonorientation.

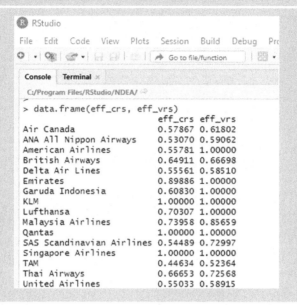

```
> data.frame(eff_crs, eff_vrs)
                             eff_crs eff_vrs
Air Canada                   0.57867 0.61802
ANA All Nippon Airways       0.53070 0.59062
American Airlines            0.55781 1.00000
British Airways              0.64911 0.66698
Delta Air Lines              0.55561 0.58510
Emirates                     0.89886 1.00000
Garuda Indonesia             0.60830 1.00000
KLM                          1.00000 1.00000
Lufthansa                    0.70307 1.00000
Malaysia Airlines            0.73958 0.85659
Qantas                       1.00000 1.00000
SAS Scandinavian Airlines    0.54489 0.72997
Singapore Airlines           1.00000 1.00000
TAM                          0.44634 0.52364
Thai Airways                 0.66653 0.72568
United Airlines              0.55033 0.58915
```

TABLE 6.7 Target and slacks inputs and outputs under CRS nonorientation Tone SBM with bad output model.

```
> data.frame(targets(result_crs), slacks(result_crs))
                          target_input.x1 target_input.x2 target_input.x3 target_output.y1 target_output.b1
Air Canada                       5019.077     11285169063        1862826.1       7577867000          5880942
ANA All Nippon Airways           4202.876      9449978938        1559894.0       6345557000          4924585
American Airlines               13435.233     30208519726        4986475.4      20284689000         15742303
British Airways                  9674.436     21401581000        3593033.5      14517800000         11343207
Delta Air Lines                 10757.897     24188649049        3992784.3      16242412000         12605220
Emirates                        11794.688     26519826443        4377588.3      17807772000         13820046
Garuda Indonesia                 1190.130      2675954444         441715.8       1796874000          1394497
KLM                              8101.000     15090771000        3027818.2      11440658000          9558822
Lufthansa                       13278.694     27007957000        4947651.4      19327674000         15619735
Malaysia Airlines                3426.498      7292543000        1274528.2       5069193000          4023686
Qantas                          12156.000     17368244000        3156052.3      12569254000          9963657
SAS Scandinavian Airlines        1754.868      3945741421         651317.7       2649522000          2056210
Singapore Airlines               9467.000     21286125000        3513669.0      14293399000         11092653
TAM                              2710.133      6093612762        1005863.6       4091794000          3175511
Thai Airways                     4558.794     10250243711        1691992.5       6882926000          5341620
United Airlines                 11250.671     25296629953        4175677.1      16986409000         13182613
                          slack_input.x1 slack_input.x2 slack_input.x3 slack_output.y1 slack_output.b1
Air Canada                     3.332923e+03     1743443937      1197944.3               0       3761910.1
ANA All Nippon Airways         2.276124e+03     5233553062       996619.8               0       3146328.7
American Airlines              9.666767e+03     4499209274      3668417.5               0      11581194.1
British Airways                6.888564e+03              0      1711378.0               0       5402820.3
Delta Air Lines                6.650103e+03     3103775951      3357162.2               0      10598560.9
Emirates                       1.358312e+03      849620557       339683.3               0       1072379.7
Garuda Indonesia               9.968698e+02      158229556       234630.7               0        740729.2
KLM                            2.353727e-11              0            0.0               0             0.0
Lufthansa                      2.000931e+04              0       812134.2               0       2563907.7
Malaysia Airlines              1.804502e+03              0       332375.8               0       1049310.5
Qantas                         0.000000e+00              0            0.0               0             0.0
SAS Scandinavian Airlines      2.291132e+03      206928579       456860.7               0       1442309.1
Singapore Airlines             0.000000e+00              0            0.0               0             0.0
TAM                            4.099867e+03     1746635238      1009232.8               0       3186147.6
Thai Airways                   2.815206e+03      190797289       725863.7               0       2291551.7
United Airlines                7.209329e+03     3768959047      3472158.2               0      10961603.5
```

TABLE 6.8 Target and slacks inputs and outputs under VRS nonorientation Tone SBM with bad output model.

```
RStudio
File   Edit   Code   View   Plots   Session   Build   Debug   Profile   Tools   Help
                                   Go to file/function           Addins

Console   Terminal
C:/Program Files/RStudio/NDEA/
> data.frame(targets(result_vrs), slacks(result_vrs))
                          target_input.x1 target_input.x2 target_input.x3 target_output.y1 target_output.b1
Air Canada                       5732.164     10181428987       2085942.7       7577867000          6585321
ANA All Nippon Airways           4836.890      9550613595       1709120.1       6345557000          5395692
American Airlines               23102.000     34707729000       8654892.9      20284689000         27323497
British Airways                 10055.399     21401581000       3697158.6      14517800000         11671930
Delta Air Lines                 11511.194     24659834524       4181167.0      16242412000         13199944
Emirates                        13153.000     27369447000       4717271.6      17807772000         14892426
Garuda Indonesia                 2187.000      2834184000        676346.5       1796874000          2135226
KLM                              8101.000     15090771000       3027818.2      11440658000          9558822
Lufthansa                       33288.000     27007957000       5759785.6      19327674000         18183643
Malaysia Airlines                4149.049      7292543000       1449802.9       5069193000          4577028
Qantas                          12156.000     17368244000       3156052.3      12569254000          9963657
SAS Scandinavian Airlines        2709.882      3917841037        884250.1       2649522000          2791578
Singapore Airlines               9467.000     21286125000       3513669.0      14293399000         11092653
TAM                              3594.348      5750869674       1235923.5       4091794000          3901810
Thai Airways                     5305.995      9298206714       1916493.2       6882926000          6050369
United Airlines                 12291.525     25947681124       4435971.2      16986409000         14004361
                         slack_input.x1  slack_input.x2  slack_input.x3  slack_output.y1 slack_output.b1
Air Canada                  2.619836e+03      2847184013         974827.6                0    3.077531e+06
ANA All Nippon Airways      1.642110e+03      5132918405         847393.7                0    2.675222e+06
American Airlines           0.000000e+00               0            0.0                0    0.000000e+00
British Airways             6.507601e+03               0        1607252.9                0    5.074097e+06
Delta Air Lines             5.896806e+03       2632590476       3168779.4                0    1.000384e+07
Emirates                    0.000000e+00               0            0.0                0    0.000000e+00
Garuda Indonesia            0.000000e+00               0            0.0                0    1.854111e-06
KLM                        -3.944155e-11               0            0.0                0    0.000000e+00
Lufthansa                   0.000000e+00               0            0.0                0   -1.280572e-06
Malaysia Airlines           1.081951e+03               0         157101.1                0    4.959681e+05
Qantas                      0.000000e+00               0            0.0                0    4.518543e-06
SAS Scandinavian Airlines   1.336118e+03       234828963         223928.2                0    7.069413e+05
Singapore Airlines          0.000000e+00               0            0.0                0    0.000000e+00
TAM                         3.215652e+03      2089378326         779172.9                0    2.459849e+06
Thai Airways                2.068005e+03      1142834286         501362.9                0    1.582803e+06
United Airlines             6.168475e+03      3117907876        3211864.1                0    1.013986e+07
>
```

that reducing bad outputs is inevitably accompanied by reduction in good outputs. Then there are times, as in the case of airlines, that a certain bad output is closely related and inseparable from a certain input. In the case of airlines, fuel burn leads to carbon emissions, meaning that to reduce carbon emissions, airlines have to reduce the amount of fuel burn. To incorporate the inseparability condition in Tone's SBM with bad output model, we refer readers to Cooper et al. (2007) Chapter 13.

6.6 Chung et al. (1997) Malmquist–Luenberger

The previous models are useful for cross-sectional data. If panel data are available, on could measure productivity change and its decompositions using the ML productivity index developed by Chung et al. (1997). ML productivity index is based on a nonparametric linear programming. It estimates the directional distance function and allows bad outputs to be incorporated into the model without the need for shadow prices. The ML index explicitly credits[4]

---

4. The term 'credit' should not be taken literally as a form of reward but rather a form of scaling between good and bad outputs as described here and in many environmental DEA studies.

firms for reductions in bad outputs, providing a measure of productivity which will inform managers whether their 'true' productivity has improved over time. In similar fashion to the Malmquist index, the ML index can be decomposed into efficiency change (catching up to the frontier) and technical change (a shift in the best practice frontier).

Maintaining our previous reasoning for adopting the input orientation, the directional input distance function for the ML index with respect to two time periods (t and $t + 1$) is defined as

$$\overrightarrow{D}_I^{t+1}(x^t, y^y, b^t : g) = \sup\left[\beta(y^t, b^t) + \beta g \in P^{t+1}(x^t)\right] \tag{6.10}$$

where x are inputs, y is the good output and b is the bad output. 'g' is the direction vector, $g = (y, -b)$, in which outputs are scaled. In the case of airline services, the production of good output, TKP, is increased, while bad output, CO_2, is decreased. β is the maximum feasible expansion of good output and contraction of bad output when the expansion and contraction are identical proportions for a given level of inputs. As the theoretical framework behind the ML index is lengthy and for the sake of brevity, we direct readers to Chung et al. (1997), Färe et al. (2001) and (2007). Directional distance function expressed in Eq. (6.10) measures observations at time t based on the technology at time $t + 1$. Hence the ML index of productivity between period t and $t + 1$ is

$$\text{ML}_t^{t+1} = \left\{ \frac{\left[1 + \overrightarrow{D}_I^{t+1}(x^t, y^t, b^t; y^t, -b^t)\right]}{\left[1 + \overrightarrow{D}_I^{t+1}\left(x^{t+1}, y^{t+1}, b^{t+1}; y^{t+1}, -b^{t+1}\right)\right]} \times \frac{\left[1 + \overrightarrow{D}_I^{t}(x^t, y^t, b^t; y^t, -b^t)\right]}{\left[1 + \overrightarrow{D}_I^{t}\left(x^{t+1}, y^{t+1}, b^{t+1}; y^{t+1}, -b^{t+1}\right)\right]} \right\}^{\frac{1}{2}}$$

$$\tag{6.11}$$

If the ML index for an airline has a value greater than 1, then the airline exhibits productivity improvement, and if the value is less than 1, there is decline in productivity. Like the Malmquist index, the ML index can also be decomposed into efficiency change (MLEFFCH) and technical change (MLTECH). This can be written as follows:

$$\text{MLEFFCH}_t^{t+1} = \left[1 + \overrightarrow{D}_I^{t}(x^t, y^t, b^t; y^t, -b^t)\right] / \left[1 + \overrightarrow{D}_I^{t+1}\left(x^{t+1}, y^{t+1}, b^{t+1}; y^{t+1}, -b^{t+1}\right)\right]$$

$$\tag{6.12}$$

$$\text{MLTECH}_t^{t+1} = \left\{ \frac{\left[1 + \overrightarrow{D}_0^{t+1}(x^t, y^t, b^t; y^t, -b^t)\right]}{\left[1 + \overrightarrow{D}_0^{t}(x^t, y^t, b^t; y^t, -b^t)\right]} \times \frac{\left[1 + \overrightarrow{D}_0^{t+1}\left(x^{t+1}, y^{t+1}, b^{t+1}; y^{t+1}, -b^{t+1}\right)\right]}{\left[1 + \overrightarrow{D}_0^{t}\left(x^{t+1}, y^{t+1}, b^{t+1}; y^{t+1}, -b^{t+1}\right)\right]} \right\}^{\frac{1}{2}}$$

$$\tag{6.13}$$

Eq. (6.12) measures the change in input efficiency between two periods. If $MLEFFCH_t^{t+1}$ exceeds 1.00, it indicates that an airline is closer to the frontier in period $t + 1$ than it was in period t. If the value is less than 1, then the airline is 'falling behind' the frontier. Eq. (6.13) measures technical change which illustrates shifts in the production possibilities frontier. If this shift is in the direction for more good output with less bad output, then the value of $MLTECH_t^{t+1}$ exceeds 1. If the value is less than 1, then technical regression has occurred. A mixed-period reference technology is also used as indicated in Eqs. (6.12) and (6.13), which is the geometric mean of two based period indices.

To calculate the ML index and its decompositions, four distance functions which are specified as LP problems must be solved. Let us assume that if at a given time $t = 1, ... T$, there are $k = 1, ... K$ airlines of inputs and outputs, the model can be expressed as

$$P(x) = \{(y, b): \sum_{k=1}^{K} z_k y_{km}^t \geq y_m^t , \ m = 1, ...M,$$

$$\sum_{k=1}^{K} z_k b_{kj}^t = b_j^t, \ j = 1, ...J,$$

$$\sum_{k=1}^{K} z_k x_{kn}^t \leq x_n^t , \ n = 1, ...N,$$

$$z_k \geq 0, k = 1, \ ... \ K\}. \tag{6.14}$$

which exhibits constant returns to scale so that:

$$P(\lambda x) = \lambda P(x), \lambda > 0 \tag{6.15}$$

and strong disposability of inputs:

$$x' \geq x \Rightarrow P(x') \supseteq P(x) \tag{6.16}$$

The inequalities for inputs and good output in Eq. (6.14) reflect the assumption that they are freely disposable. The bad output is assumed to be costly to dispose of and, therefore, are modelled as equalities (i.e. weak disposable). The nonnegativity constraints on the intensity variables, z_k, allow the model to exhibit constant returns to scale.[5]

---

5. CRS is used because it is a necessary condition for the resulting productivity indices to be a true total factor productivity index (Färe and Grosskopf, 1996). The rationale for constant returns to scale is because it is consistent with the vast majority of airline literature such as White (1979), Cornwell et al. (1990), Good et al. (1995), Alam and Sickles (2000) and Sickles et al. (2002). As empirically demonstrated in Caves et al. (1984), large and small carriers in the United States could compete with one another over extended periods of time. They found that airline production technology can be characterized as having constant returns to scale.

6.6.1 R script for Malmquist−Luenberger model

The R script for the ML model is shown in Table 6.9.

TABLE 6.9 R script for Malmquist−Luenberger model.

```
D <- read.table(file="C:/Airline/Delivery_ML.txt",header=TRUE)
attach(D)

X1 <- D[D[,"Year"]==1,c("x1","x2","x3")]
X2 <- D[D[,"Year"]==2,c("x1","x2","x3")]

Y1 <- as.matrix(D[D[,"Year"]==1,"y1"])
Y2 <- as.matrix(D[D[,"Year"]==2,"y1"])

U1 <- as.matrix(D[D[,"Year"]==1,"b1"])
U2 <- as.matrix(D[D[,"Year"]==2,"b1"])

#DDF
library(lpSolve)

ddf <- function(X,Y,U,XREF=X,YREF=Y,UREF=U,gy=1,gu=-1,K=2){

  n <- nrow(X); m <- ncol(X); k <- ncol(Y); s <- ncol(U); l <- nrow(XREF)
  beta <- c(); lambda <- c()
  results <- t(sapply(1:n,function(i){
          ZF <- c(1,-1,rep(0,times=l))
          NB_LHS <- rbind(cbind(0,0,-t(XREF)),cbind(gy*t(Y)[,i],-gy*t(Y)[,i],-
t(YREF)),cbind(gu*t(U)[,i],-gu*t(U)[,i],-t(UREF)),c(0,0,rep(1,times=l)))
          NB_RHS <- c(-t(X)[,i],-t(Y)[,i],-t(U)[,i],0)
          NB_Dir <- c(rep(">=",times=m),rep("<=",times=k),rep(">=",times=s),">=")
          LP <- lp("max",ZF,NB_LHS,NB_Dir,NB_RHS)
          if(LP$status==2){beta <- c(beta,NA); lambda <- rbind(lambda,LP$solution[-
c(1:2)])}
                  else {beta <- c(beta,LP$solution[1]-LP$solution[2]); lambda <-
rbind(lambda,LP$solution[-c(1:2)])}
                  return(c(beta,lambda))
                  }))
  colnames(results) <- c("beta",paste("lambda",1:l))
  return(list(beta=results[,"beta"],lambda=results[,-1]))
                  }

#MALMQUIST
malmquist <- function(dist11,dist12,dist21,dist22){
                  effch <- dist11/dist22
                  techch <- as.vector(sqrt((dist12/dist11)*(dist22/dist21)))
                  prodch <- effch*techch
                  return(list(effch=effch,techch=techch,prodch=prodch))
                  }

                  dist11 <- 1+ddf(X=X1,Y=Y1,U=U1,XREF=X1,YREF=Y1,UREF=U1)$beta
                  dist22 <- 1+ddf(X=X2,Y=Y2,U=U2,XREF=X2,YREF=Y2,UREF=U2)$beta
                  dist12 <- 1+ddf(X=X1,Y=Y1,U=U1,XREF=X2,YREF=Y2,UREF=U2)$beta
                  dist21 <- 1+ddf(X=X2,Y=Y2,U=U2,XREF=X1,YREF=Y1,UREF=U1)$beta
                  malm <- malmquist(dist11,dist12,dist21,dist22)
                  prodch <- malm$prodch; effch <- malm$effch; techch <- malm$techch

data.frame(prodch,effch,techch)
data.frame(dist11, dist22, dist12, dist21)
```

> ">=" for s refers to
> weak disposability.
> ">=" refers to CRS.

> Refers to Equations
> 6.12 and 6.13,
> respectively.

6.6.2 Interpretation of Malmquist–Luenberger results

Table 6.10 present the ML productivity change estimates (prodch) and its decomposition into ML effch and ML techch from 2008 to 2011 using the data set presented in Appendix D.3. The years 1, 2, 3 and 4 correspond to the years 2008, 2009, 2010 and 2011. The dmus F1, F2 ⋯ F16 aligns with the airlines shown in Appendix D.1. It is worth noting that ML sometimes results in infeasible LP problem as observed in Fare et al. (2001), Aparicio et al. (2013) and Du et al. (2018).

The results from Table 6.10 show that productivity gains are attributed to efficiency change suggesting that airlines, taking into carbon emissions, have adopted best practice management measures. *Techch* for all airlines was negative (i.e. below unity), which suggests that airlines have yet to adopt new

TABLE 6.10 Productivity change, efficiency change and technical change of ML.

```
R  RStudio

File   Edit   Code   View   Plots   Session   Buil

 O  ▾  ▱  ▱ ▾  ▯ ▯  ▯  ▸ Go to file/func

Console   Terminal

C:/Program Files/RStudio/NDEA/
> data.frame(prodch,effch,techch)
          prodch       effch      techch
1    1.0015475  1.0049076  0.9966563
2    0.9933103  0.9999201  0.9933896
3    0.9954947  1.0009242  0.9945755
4    0.9979303  1.0106411  0.9874231
5    1.0004643  1.0103761  0.9901900
6    0.9657577  0.9738544  0.9916859
7    0.9400513  0.9508313  0.9886625
8    0.9747069  1.0000000  0.9747069
9    0.9801359  0.9923569  0.9876849
10   1.0032117  1.0142612  0.9891059
11   1.0078063  1.0154339  0.9924883
12   0.9508608  0.9629720  0.9874231
13   0.9636740  1.0000000  0.9636740
14   0.9961946  1.0004398  0.9957567
15   0.9693851  0.9814924  0.9876644
16   0.9927789  1.0002458  0.9925349
```

technologies that may further increase the good output while lowering CO_2 levels. It is also important to emphasize here that our sample size is small and, with a short time series, that it is unlikely that any gains in technical change can be realized.

When compared with Table 4.10 based on Malmquist productivity index, of the seven airlines that exhibited productivity growth, Table 6.10 reveal only four airlines exhibited productivity growth—Air Canada, Delta Air lines, Malaysia Airlines and Qanats. What this suggests is that when carbon emission is taken into account in the production model, the model internalizes pollution, which thus credits a firm for simultaneously reducing production of bad output (carbon emissions) and increasing production of good outputs (passenger, freight and mail).

6.7 Conclusions

In this chapter, we covered some models that incorporated bad outputs, namely, Seiford and Zhu (2002) transformation approach, Zhou et al. (2008) radial efficiency measure, Tone's SBM approach with bad outputs, and Chung et al. (1997) ML productivity index. The main purpose was to provide a range of models so that readers can appreciate the complexities involved in modelling DEA to incorporate bad outputs.

The first three models, which are only used for cross-sectional analysis, revealed consistency in the results under VRS, that the same seven airlines were identified as efficient. Under CRS, both Zhou et al. (2008) radial efficiency measure and Tone's SBM approach with bad outputs revealed the same three airlines as efficient. The outcomes from these models validate their robustness and theoretical foundation. Nonetheless, the issue regarding the choice of input or output orientation has been a contentious one albeit a minor subject in measuring efficiency. As noted in Cooper et al. (2007), only the nonradial and nonoriented models can capture the whole aspects of efficiency, suggesting that the nonseparable (NS) SBM model and NS-Overall models can deal with inputs and outputs efficiency in the presence of both inseparable and separable outputs in a unified manner in any returns-to-scale environment. With regards to the ML index, this model has been used widely across a range of environmental studies such as Färe et al. (2001) on US state manufacturing, Kumar (2006) on economy wide global analysis, and Du et al. (2018) on China's environmental performance. But as argued in O'Donnell (2010) that the MPI, including MLI as it is derived from MPI, is multiplicatively incomplete and fails to capture productivity changes associated with changes in scope (i.e. changes in output mix and input mix). To overcome these issues, one could use Dakpo et al. (2019) pollution-adjusted Färe—Primont productivity index.

Appendix D

TABLE D.1 Dataset for 'Delivery_bad.txt', 2009

DMU	Flight personnel x1	Available tonne kilometres (thousands) x2	Fuel burn (tonnes) x3	Passenger, freight and mail tonne kilometres performed (thousands) y1	CO_2 emissions (tonnes) b1
Air Canada	8352	13,028,613	3,060,770.35	7,577,867	9,662,852
All Nippon Airways (ANAs)	6479	14,683,532	2,556,513.78	6,345,557	8,070,914
American Airlines	23,102	34,707,729	8,654,892.94	20,284,689	27,323,497
British Airways	16,563	21,401,581	5,304,411.47	14,517,800	16,746,027
Delta Air Lines	17,408	27,292,425	7,349,946.47	16,242,412	23,203,781
Emirates	13,153	27,369,447	4,717,271.61	17,807,772	14,892,426
Garuda Indonesia	2187	2,834,184	676,346.53	1,796,874	2,135,226
KLM	8101	15,090,771	3,027,818.18	11,440,658	9,558,822
Lufthansa	33,288	27,007,957	5,759,785.56	19,327,674	18,183,643
Malaysia Airlines	5231	7,292,543	1,606,904.02	5,069,193	5,072,996
Qantas	12,156	17,368,244	3,156,052.26	12,569,254	9,963,657
SAS Scandinavian Airlines	4046	4,152,670	1,108,178.33	2,649,522	3,498,519
Singapore Airlines	9467	21,286,125	3,513,668.99	14,293,399	11,092,653
TAM	6810	7,840,248	2,015,096.39	4,091,794	6,361,659
Thai Airways	7374	10,441,041	2,417,856.19	6,882,926	7,633,172
United Airlines	18,460	29,065,589	7,647,835.29	16,986,409	24,144,216

TABLE D.2 Results of technical efficiency (CRS) output orientation using data set of Appendix D.1 (excluding b1).

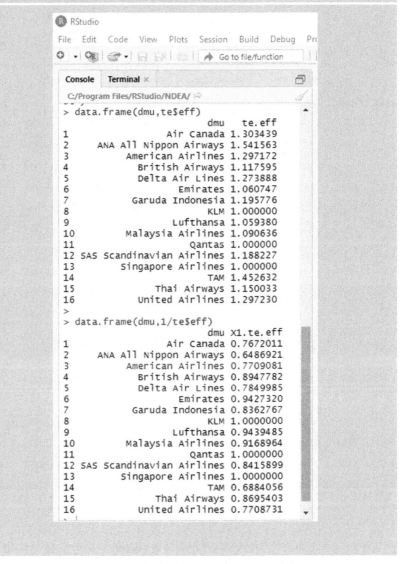

```
R  RStudio
File   Edit   Code   View   Plots   Session   Build   Debug   Pr

      Go to file/function

Console   Terminal ×

C:/Program Files/RStudio/NDEA/
> data.frame(dmu,te$eff)
                         dmu   te.eff
1                  Air Canada 1.303439
2       ANA All Nippon Airways 1.541563
3           American Airlines 1.297172
4             British Airways 1.117595
5             Delta Air Lines 1.273888
6                    Emirates 1.060747
7            Garuda Indonesia 1.195776
8                         KLM 1.000000
9                   Lufthansa 1.059380
10           Malaysia Airlines 1.090636
11                     Qantas 1.000000
12 SAS Scandinavian Airlines 1.188227
13          Singapore Airlines 1.000000
14                         TAM 1.452632
15                Thai Airways 1.150033
16             United Airlines 1.297230
>
> data.frame(dmu,1/te$eff)
                         dmu   X1.te.eff
1                  Air Canada 0.7672011
2       ANA All Nippon Airways 0.6486921
3           American Airlines 0.7709081
4             British Airways 0.8947782
5             Delta Air Lines 0.7849985
6                    Emirates 0.9427320
7            Garuda Indonesia 0.8362767
8                         KLM 1.0000000
9                   Lufthansa 0.9439485
10           Malaysia Airlines 0.9168964
11                     Qantas 1.0000000
12 SAS Scandinavian Airlines 0.8415899
13          Singapore Airlines 1.0000000
14                         TAM 0.6884056
15                Thai Airways 0.8695403
16             United Airlines 0.7708731
```

TABLE D.3 Dataset for 'Delivery_ML.txt'.

Year	dmu	Flight personnel	Available tonne kilometres (thousands)	Fuel burn (tonnes)	Passenger, freight and mail tonne kilometres performed (thousands)	CO_2 emissions (tonnes)
		x1	x2	x3	y1	b1
1	F1	9143	13,607,675,000	3,237,847	7,963,161,000	10,221,882
1	F2	6230	15,599,193,000	2,717,899	6,866,683,000	8,580,408
1	F3	24,267	37,407,051,000	9,406,144	22,159,849,000	29,695,197
1	F4	18,390	22,398,308,000	5,599,238	15,225,443,000	17,676,793
1	F5	18,267	28,866,505,000	7,698,555	17,067,385,000	24,304,340
1	F6	13,442	23,752,982,000	3,881,635	15,764,086,000	12,254,321
1	F7	2287	2,575,254,000	677,726	1,760,887,000	2,139,582
1	F8	8433	16,336,676,000	3,214,711	12,542,959,000	10,148,843
1	F9	27,376	29,350,489,000	5,988,763	21,193,962,000	18,906,526
1	F10	5377	8,457,146,000	1,795,361	5,727,371,000	5,667,956
1	F11	11,475	15,484,945,000	3,368,147	11,708,313,000	10,633,239
1	F12	4553	4,971,277,000	1,354,921	3,335,732,000	4,277,485
1	F13	10,234	24,362,864,000	3,984,628	16,495,404,000	12,579,471
1	F14	7069	7,034,568,000	1,789,062	3,677,442,000	5,648,069
1	F15	7430	10,861,259,000	2,555,078	7,429,327,000	8,066,383
1	F16	19,517	31,987,031,000	8,315,963	18,836,162,000	26,253,495
2	F1	8352	13,028,613,000	3,060,770	7,577,867,000	9,662,852
2	F2	6479	14,683,532,000	2,556,514	6,345,557,000	8,070,914
2	F3	23,102	34,707,729,000	8,654,893	20,284,689,000	27,323,497
2	F4	16,563	21,401,581,000	5,304,411	14,517,800,000	16,746,027
2	F5	17,408	27,292,425,000	7,349,946	16,242,412,000	23,203,781
2	F6	13,153	27,369,447,000	4,717,272	17,807,772,000	14,892,426
2	F7	2187	2,834,184,000	676,347	1,796,874,000	2,135,226
2	F8	8101	15,090,771,000	3,027,818	11,440,658,000	9,558,822

TABLE D.3 Dataset for 'Delivery_ML.txt'.—cont'd

Year	dmu	Flight personnel	Available tonne kilometres (thousands)	Fuel burn (tonnes)	Passenger, freight and mail tonne kilometres performed (thousands)	CO$_2$ emissions (tonnes)
		x1	x2	x3	y1	b1
2	F9	33,288	27,007,957,000	5,759,786	19,327,674,000	18,183,643
2	F10	5231	7,292,543,000	1,606,904	5,069,193,000	5,072,996
2	F11	12,156	17,368,244,000	3,156,052	12,569,254,000	9,963,657
2	F12	4046	4,152,670,000	1,108,178	2,649,522,000	3,498,519
2	F13	9467	21,286,125,000	3,513,669	14,293,399,000	11,092,653
2	F14	6810	7,840,248,000	2,015,096	4,091,794,000	6,361,659
2	F15	7374	10,441,041,000	2,417,856	6,882,926,000	7,633,172
2	F16	18,460	29,065,589,000	7,647,835	16,986,409,000	24,144,216
3	F1	8569	13,958,344,000	3,219,901	8,557,102,000	10,165,228
3	F2	6214	14,740,048,000	2,578,215	7,053,773,000	8,139,424
3	F3	22,843	34,768,045,000	8,643,187	21,069,160,000	27,286,542
3	F4	16,656	20,844,337,000	5,181,557	14,078,536,000	16,358,177
3	F5	28,094	44,162,438,000	11,443,559	27,606,002,000	36,127,316
3	F6	14,968	31,395,942,000	5,433,732	21,779,289,000	17,154,291
3	F7	2662	3,585,816,000	760,580	2,197,597,000	2,401,151
3	F8	8322	14,437,621,000	3,072,141	11,408,232,000	9,698,748
3	F9	34,319	27,195,318,000	5,896,808	20,744,057,000	18,616,222
3	F10	5172	7,780,030,000	1,621,842	5,889,318,000	5,120,156
3	F11	12,837	18,200,001,000	3,119,716	14,936,911,000	9,848,942
3	F12	3739	3,986,369,000	1,056,254	2,861,152,000	3,334,595
3	F13	9857	21,535,099,000	3,568,900	15,103,426,000	11,267,016
3	F14	8517	8,701,273,000	2,147,523	4,785,087,000	6,779,730
3	F15	7209	11,478,797,000	2,503,521	7,983,169,000	7,903,616
3	F16	18,327	28,975,323,000	7,735,753	17,706,994,000	24,421,773

Continued

TABLE D.3 Dataset for 'Delivery_ML.txt'.—cont'd

Year	dmu	Flight personnel x1	Available tonne kilometres (thousands) x2	Fuel burn (tonnes) x3	Passenger, freight and mail tonne kilometres performed (thousands) y1	CO_2 emissions (tonnes) b1
4	F1	8700	14,578,014,000	3,327,498	8,801,111,000	10,504,911
4	F2	6358	16,011,266,000	2,820,653	7,374,217,000	8,904,801
4	F3	23,075	34,780,733,000	8,705,656	21,065,417,000	27,483,755
4	F4	17,686	22,743,599,000	5,522,346	15,299,710,000	17,434,047
4	F5	28,706	44,955,715,000	11,744,048	27,735,644,000	37,075,958
4	F6	17,026	34,340,999,000	5,949,734	22,789,814,000	18,783,311
4	F7	3025	4,564,624,000	940,763	2,788,371,000	2,969,988
4	F8	8791	15,409,380,000	3,278,757	12,018,549,000	10,351,037
4	F9	33,603	29,944,243,000	6,449,860	22,161,987,000	20,362,209
4	F10	5231	7,773,334,000	1,693,755	5,671,942,000	5,347,184
4	F11	12,541	18,779,627,000	3,212,271	16,075,004,000	10,141,141
4	F12	3805	4,252,768,000	1,136,306	2,981,675,000	3,587,317
4	F13	9996	22,579,442,000	3,843,804	15,380,315,000	12,134,888
4	F14	8616	9,543,028,000	2,309,737	6,021,704,000	7,291,841
4	F15	7409	11,939,041,000	2,572,847	7,819,082,000	8,122,478
4	F16	18,188	28,302,694,000	7,758,080	17,045,070,000	24,492,258

References

Alam, I.M.S., Sickles, R.C., 2000. Time series analysis of deregulatory dynamics and technical efficiency: the case of the U.S. airline industry. International Economic Review 41, 203–218.

Ali, A.I., Seiford, L.M., 1990. Translation invariance in data envelopment analysis. Operations Research Letters 9, 403–405.

Aparicio, J., Pastor, J.T., Zofio, J.L., 2013. On the inconsistency of the Malmquist–Luenberger index. European Journal of Operational Research 229 (3), 738–742.

Banker, R.D., Charnes, A., Cooper, W.W., 1984. Some models for estimating technical and scale inefficiencies in data envelopment analysis. Management Science 30, 1078–1092.

Berg, S.A., Forsund, F.R., Jansen, E.S., 1992. Malmquist indices of productivity growth during the deregulation of Norwegian banking 1980–1989. The Scandinavian Journal of Economics 94, 211–228 (Supplement).

Caves, D.W., Christensen, L.R., Tretheway, M.W., 1984. Economies of density vs. economies of scale: why trunk and local service airline costs differ. The RAND Journal of Economics 15, 471–489.

Charnes, A., Cooper, W.W., 1962. Programming with linear fractional functionals. Naval Research Logistics Quarterly 15, 333–334.

Chung, Y.H., Färe, R., Grosskopf, S., 1997. Productivity and undesirable outputs: a directional distance function approach. Journal of Environmental Management 51, 229–240.

Coelli, T., Rao, D.S.P., Battese, G.E., 1998. An Introduction to Efficiency and Productivity Analysis. Kluwer Academic Publishers, Boston, MA.

Coll-Serrano, V., Bolós, V., Benítez, R., 2018. Conventional and Fuzzy Data Envelopment Analysis. R Package 'deaR'. Version 1.3. https://cran.r-project.org/web/packages/deaR/deaR.pdf.

Cooper, W.W., Seiford, L.M., Tone, K., 2007. Data Envelopment Analysis A Comprehensive Text with Models, Applications, References and DEA-Solver Software, second ed. Springer, New York.

Cornwell, C., Schmidt, P., Sickles, R.C., 1990. Production frontiers with cross sectional and time-series variation in efficiency levels. Journal of Econometrics 46, 185–200.

Cui, Q., Wei, Y.M., Li, Y., 2016. Exploring the impacts of the EU ETS emission limits on airline performance via the dynamic environmental DEA approach. Applied Energy 183, 984–994.

Dakpo, K.H., Jeanneaux, P., Latruffe, L., 2019. Pollution-adjusted productivity changes: extending the färe–primont index with an illustration with French suckler cow farms. Environmental Modeling and Assessment 24, 625–639. https://doi.org/10.1007/s10666-019-09656-y.

Du, J., Chen, Y., Huang, Y., 2018. A modified malmquist-luenberger productivity index: assessing environmental productivity performance in China. European Journal of Operational Research 269, 171–187.

Du, J., Duan, Y.-R., Xu, J.-H., 2019. The infeasible problem of Malmquist–Luenberger index and its application on China's environmental total factor productivity. Annals of Operations Research 278, 235–253.

Fare, R., Grosskopf, S., Lovell, C.A.K., Yaiswarng, S., 1993. Deviation of shadow prices for undesirable outputs: a distance function approach. The Review of Economics and Statistics 75 (2), 374–380.

Färe, R., Grosskopf, S., 1996. Intertemporal Production Frontiers: With Dynamic DEA. Kluwer Academic, Boston, MA.

Färe, R., Grosskopf, S., Sickles, R.C., 2007. Productivity of US airlines after deregulation. Journal of Transport Economics and Policy 40, 93–112.

Färe, R., Grosskopf, S., 2004. Modeling undesirable factors in efficiency evaluation: comment. European Journal of Operational Research 157, 242–245.

Färe, R., Grosskopf, S., Lovell, C.A.K., 1994. Production Frontiers. Cambridge University Press, Cambridge.

Färe, R., Grosskopf, S., Lovell, C.A.K., Pasurka, C., 1989. Multilateral productivity comparisons when some outputs are undesirable: a nonparametric approach. The Review of Economics and Statistics 71, 90–98.

Färe, R., Grosskopf, S., Pasurka, C., 2001. Accounting for air pollution emissions in measures of state manufacturing productivity growth. Journal of Regional Science 41, 381–409.

Good, D.H., Roller, L.H., Sickles, R.C., 1995. Airline efficiency differences between Europe and the US: implications for the pace of EC integration and domestic regulation. European Journal of Operational Research 80, 508–518.

Hailu, A., Veeman, T., 2001. Non-parametric productivity analysis with undesirable outputs: an application to Canadian pulp and paper industry. American Journal of Agricultural Economics 83 (3), 605–616.

Halkos, G., Petrou, K.N., 2019. Treating undesirable outputs in DEA: a critical review. Economic Analysis and Policy 62, 97–104.

Hua, Z., Bian, Y., Liang, L., 2007. Eco-efficiency analysis of paper mills along the Huai River: an extended DEA approach. Omega 35 (5), 578–587.

Koopmans, T.C., 1951. Analysis of production as an efficient combination of activities. In: Koopmans, T.C. (Ed.), Activity Analysis of Production and Allocation, Cowles Commission for Research in Economics. Wiley, New York, pp. 33–97.

Korhonen, P., Luptacik, M., 2004. Eco-efficiency analysis of power plants: an extension of data envelopment analysis. European Journal of Operational Research 154, 437–446.

Kumar, S., 2006. Environmentally sensitive productivity growth: a global analysis using Malmquist–Luenberger index. Ecological Economics 56, 280–293.

Lee, B.L., Wilson, C., Pasurka, C., 2015. The good, the bad and the efficient: productivity, efficiency and technical change in the Airline Industry, 2004–2011. Journal of Transport Economics and Policy 49, 338–354.

Lee, B.L., Wilson, C., Pasurka, C.A., Fujii, H., Managi, S., 2017. Sources of airline productivity from carbon emissions: an analysis of operational performance under good and bad outputs. Journal of Productivity Analysis 47 (3), 223–246.

Liu, X., Zhou, D., Zhou, P., Wang, Q.-W., 2017. Dynamic carbon emission performance of Chinese airlines: a global Malmquist index analysis. Journal of Air Transport Management 65, 99–109.

O'Donnell, C.J., 2010. Measuring and decomposing agricultural productivity and profitability change. The Australian Journal of Agricultural and Resource Economics 54, 527–560.

Reinhard, S., Lovell, C.A.K., Thijssen, G., 1999. Econometric estimation of technical and environmental efficiency: an application to Dutch dairy farms. American Journal of Agricultural Economics 81, 44–60.

Scheel, H., 2001. Undesirable outputs in efficiency valuations. European Journal of Operational Research 132, 400–410.

Seiford, L.M., Zhu, J., 2002. Modeling undesirable factors in efficiency evaluation. European Journal of Operational Research 142 (1), 16–20.

Sickles, R.C., Good, D.H., Getachew, L., 2002. Specification of distance functions using semi- and nonparametric methods with an application to the dynamic performance of eastern and western European air carriers. Journal of Productivity Analysis 17, 133–155.

Tone, K., 2001. A slacks-based measure of efficiency in data envelopment analysis. European Journal of Operational Research 130 (3), 498–509.

Tone, K., Tsutsui, M., 2011. Applying an efficiency measure of desirable and undesirable outputs in DEA to U.S. Electric utilities. Journal of Centrum Cathedra 4, 236–249.

White, L.J., 1979. Economies of scale and the question of natural monopoly in the airline industry. Journal of Air Law and Commerce 44, 545–573.

Yaisawarng, S., Klein, J.D., 1994. The effects of sulfur dioxide controls on productivity change in the US electric power industry. The Review of Economics and Statistics 76, 447–460.

Zhou, P., Ang, B.W., Poh, K.L., 2008. Measuring environmental performance under different environmental dea technologies. Energy Economics 30 (1), 1–14.

Zofio, J.L., Prieto, A.M., 2001. Environmental efficiency and regulatory standards: the case of CO_2 emissions from OECD industries. Resource and Energy Economics 23, 63–83.

Chapter 7

Measuring airline performance: Network DEA

7.1 Introduction

In Chapter 3, we described the standard measure of DEA using the *'provision'* and *'delivery'* models separately. In the *provision* model, we considered two inputs—the number of maintenance employees and value of assets (property, plant and equipment), which produces one output, ATK. In the *delivery* model, ATK is an input and, together with fuel burn and flight personnel, produces outputs passenger TKP and freight and mail TKP. This description of an airline operation is analogous to a supply chain management, and due to the multiple divisions within its operations, the standard DEA would not be an appropriate model for measuring airline efficiency. As argued in Chapter 3, when measuring airline performance, it is important to accurately define the production function such that the inputs used align with the outputs produced. It is widely recognized that standard DEA treats the production process as a 'black box' in that it simply transforms inputs into outputs and neglects any possible intervening processes, including dissimilar series or parallel functions. As noted in Färe and Grosskopf (2000), Castelli et al. (2004) and Tone and Tsutsui (2009), DMUs may have internal or network structures, and by not appropriately identifying the production process would result in mismeasuring the relative performance of a DMU. Standard DEA also does not provide adequate information regarding sources of inefficiency within divisions of a DMU and fails to account for the continuity of links between divisions.

To accurately capture the production process and account for divisional efficiencies as well as the overall efficiency in a unified framework, we rely on the network DEA model. Network DEA (NDEA) model was first introduced in Färe (1991) and further developed in Färe and Grosskopf (1996) and Färe and Grosskopf (2000). They were the first to investigate the so-called 'black box', and since then, their NDEA model has been extended by numerous authors as described in Halkos et al. (2014). Classified supply-chain models into four categories—independent, connected, relational and game theory, which

Productivity and Efficiency Measurement of Airlines. https://doi.org/10.1016/B978-0-12-812696-7.00002-7

provide decision-makers useful information about the suitability and appropriateness of each model. As the focus of this book is on measuring airline performance, the approach that is most closely aligned with is the 'connected' approach as airlines do consider the interactions between the nodes. It also aligns with the 'relational' approach as it takes into account the mathematical relationship that exists between the nodes.

In this chapter, we demonstrate select NDEA models by combining the two models '*provision*' and '*delivery*' into one model—*provision—delivery* model. We use the 2009 data from Chapter 3 Appendix A Tables A.1 and A.3 and, set up our NDEA data set 'NDEA1' in Appendix E Tables E.1 and E.2 and apply it to the NDEA models that fall under the 'connected' and 'relational' approaches in this chapter. Additional data are also included in Appendix E Tables E.3 and E.5 based on the NDEA models used. These are described in the following sections.

This chapter is divided into 10 sections. Following the introduction in Section 7.1, we describe the basic two-node network DEA. In Section 7.3, we describe the Kao and Hwang (2008) and Liang et al. (2008) centralized model. Section 7.4 describes the network technical Farrell efficiency model. Section 7.5 describes the network cost efficiency model. Section 7.6 describes the network revenue efficiency model. Section 7.7 describes the NDEA directional distance function model. Section 7.8 describes the network slacks-based inefficiency model. Section 7.9 describes the general network technology model based on the slacks-based inefficiency model to depict the airline *provision—delivery* model. This chapter concludes with some brief comments in Section 7.10.

7.2 A basic two-node network DEA

We start our NDEA model by using a basic two-node model as depicted in Fig. 7.1.[1]

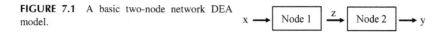

FIGURE 7.1 A basic two-node network DEA model.

---

1. For clarity purpose, we use the term 'node' instead of 'stage'. While many studies use the term 'stage', which is perfectly fine, we use 'node' to differentiate from Chapter 8, when the term 'stage' in second-stage regression is used.

The model in Fig. 7.1 assumes a DMU employing inputs $x = (x_1, \ldots, x_N)$ in node 1 to produce intermediate product $z = (z_1, \ldots, z_Q)$, which is then used in node 2 to produce final output $y = (y_1, \ldots, y_M)$. In other words,

$$\text{Network technology} = \left\{ (x,y) \left| \begin{array}{ll} x \ \varepsilon\Re_+^N \text{ can produce } z \ \varepsilon\Re_+^Q & \text{(Node 1)} \\ z \ \varepsilon\Re_+^Q \text{ can produce } y \ \varepsilon\Re_+^M & \text{(Node 2)} \end{array} \right. \right\}$$

$$\text{inputs } X = (x_1, \ldots, x_n, \ldots, x_N)^T \varepsilon\Re_+^N$$

$$\text{outputs } Y = (y_1, \ldots, y_m, \ldots, y_M)^T \varepsilon\Re_+^M$$

$$\text{intermediate products } Z = (z_1, \ldots, z_q, \ldots, z_Q)^T \varepsilon\Re_+^Q$$

The model assumes that the intermediate measures are the only inputs to the second stage, i.e. that there are no additional independent inputs to that node.

7.3 Kao and Hwang (2008) and Liang et al. (2008) network DEA centralized model

Kao and Hwang (2008) and Liang et al. (2008) extended the standard radial DEA model into a cooperative game approach (i.e. centralized model) and derived efficiency scores for both individual nodes, and the product of the two efficiency scores results in the overall efficiency score.[2]

Following the production model of Fig. 7.1, we assume DMU_j ($j = 1, 2, \ldots, n$) has Q intermediate measures, z_{qj} ($q = 1, 2, \ldots, Q$). For DMU_j, we denote the efficiency for node 1 as e_j^1 and node 2 as e_j^2. Based upon the radial CRS DEA model of Charnes et al. (1978), we define

$$e_j^1 = \sum_{q=1}^{Q} w_q z_{qj} \bigg/ \sum_{i=1}^{m} v_i x_{ij}$$

and

$$e_j^2 = \sum_{r=1}^{s} u_r y_{rj} \bigg/ \sum_{q=1}^{S} \widetilde{w}_q z_{qj} \qquad (7.1)$$

where v_i, w_q, $\widetilde{w}_q$ and u_r are unknown nonnegative weights. Note that w_q is set

---

2. Liang et al. (2008) two-stage model also includes the Stackelberg game (i.e. noncooperative game) whereby one stage is a leader and the other stage is a follower. Due to the nature of the model, which does not reflect the production process of the airline 'provision—delivery' model, we do not include the Stackelberg game here.

equal to $\tilde{w}_q$ in Eq. (7.1) as this is a characteristic of the centralized model. This means that

$$e_j^1 \times e_j^2 = \sum_{r=1}^{s} u_r y_{rj} / \sum_{i=1}^{m} v_i x_{ij} \tag{7.2}$$

Hence, the centralized model can be expressed as

$$e_j^{\text{centralized}} = \text{Max } e_j^1 \times e_j^2 = \sum_{r=1}^{s} u_r y_{rj} / \sum_{i=1}^{m} v_i x_{ij} \tag{7.3}$$

s.t.

$$e_j^1 \leq 1, \ e_j^2 \leq 1 \text{ and } w_q = \tilde{w}_q$$

Eq. (7.3) can then be converted into the following linear program:

$$e_j^{\text{centralized}} = \sum_{r=1}^{s} u_r y_{rj} / \sum_{i=1}^{m} v_i x_{ij}$$

s.t.

$$\sum_{r=1}^{s} u_r y_{rj} - \sum_{q=1}^{Q} w_q z_{qj} \leq 0, \quad j = 1, 2, \ldots, n,$$

$$\sum_{q=1}^{Q} w_q z_{qj} - \sum_{i=1}^{m} v_i x_{ij} \leq 0, \quad j = 1, 2, \ldots, n,$$

$$\sum_{i=1}^{m} v_i x_{ij} = 1$$

$$w_q \geq 0, \ q = 1, 2, \ldots, Q; \ v_i \geq 0, \ i = 1, 2, \ldots, m; \ u_r \geq 0, \ r = 1, 2, \ldots, s. \tag{7.4}$$

Eq. (7.4) yields the overall efficiency of the two-stage process. Assume that Eq. (7.4) provides a unique solution. Then the efficiencies for Node 1 and Node 2 are

$$e_j^{1, \text{ centralized}} = \sum_{q=1}^{Q} w_q^* z_{qj} / \sum_{i=1}^{m} v_i^* x_{ij} = \sum_{q=1}^{Q} w_q^* z_{qj}$$

and

$$e_j^{2, \text{ centralized}} = \sum_{r=1}^{s} u_r^* y_{rj} / \sum_{q=1}^{Q} w_q^* z_{qj} \tag{7.5}$$

If we denote the optimal value to Eq. (7.4) as $e_j^{\text{centralized}}$, then we have an efficiency decomposition of $e_j^{\text{centralized}} = e_j^{1, \text{ centralized}} \times e_j^{2, \text{ centralized}}$, as also proposed by Kao and Hwang (2008). In essence, under input-oriented DEA model, $e_j^{1,\text{centralized}} \leq 1$ and $e_j^{2,\text{centralized}} \leq 1$. This suggests that the two-stage model is efficient, and only if $e_j^{1,\text{centralized}} = e_j^{2,\text{centralized}} = 1$.

7.3.1 R script for Kao and Hwang (2008) and Liang et al. (2008)

The R script to derive the efficiency scores for node 1, 2 and the centralized model is based on 'DJL' package version 3.7 developed by Lim (2021) as shown in Table 7.1.[3]

It is, however, important to note that Kao and Hwang (2008) and Liang et al. (2008) model considers intermediate product (z) that are produced in 1 using input X. When considering the data shown in Appendix E Table E.1, it is unrealistic to say that inputs 'number of maintenance employees' and 'assets' produce 'flight personnel' and 'fuel burn'. 'Flight personnel' and 'fuel burn' are inputs that are determined by demand. Nonetheless, for the sake of maintaining consistency in the data used between chapters, we shall assume that 'flight personnel' and 'fuel burn' are generated from the use of 'number of maintenance employees' and 'assets' because of derived demand. Later on in Section 7.9, we will modify the NDEA model to include inputs that are

TABLE 7.1 R script for Kao and Hwang (2008) and Liang et al. (2008) NDEA centralized model.

```
library(DJL)
## Load data (Reading from C drive and folder of data file)
data <- read.table(file="C:/Airline/NDEA1.txt", header=T,sep="\t")
attach(data)
X = cbind(x1,x2)
Y = cbind(y1,y2)
Z = cbind(z1,z2,z3)

## Centralized (Cooperative game)
result.coop <- dm.network.dea(xdata.s1 = X, zdata = Z, ydata.s2 = Y, type = "co", rts="crs",
orientation="i")

Eff_cent_overall<- result.coop$eff.s1 * result.coop$eff.s2

data.frame(Eff_cent.node1= result.coop$eff.s1, Eff_cent.node2=
result.coop$eff.s2,Eff_cent_overall)
```

---

3. 'DJL' version 3.7 is in an alpha test stage. As such, the output orientation mode is unavailable.

directly employed only in node 2 (unrelated to Node 1) to produce the final output to depict an accurate picture of the *provision—delivery* model.

7.3.2 Interpretation of results

The efficiency scores for node 1, 2 and the NDEA centralized model are shown in Table 7.2.

The results from Table 7.2 show Singapore Airlines as the sole efficient airline. We observe that some airlines are efficient either nodes, such as Garuda Indonesia and United Airlines in 1, and KLM and Qantas in 2 suggesting that overall efficiency depends on performance of both nodes. While the results stress the significance of the impact each node's performance has on the overall performance of an airline, we should be mindful that the aforementioned results may not be meaningful as some inputs should not be considered as intermediates as explained in the previous section.

One could also derive the scale efficiency (as described in Chapter 3) by taking the ratio of the CRS efficiency to the VRS efficiency as discussed in Kao and Hwang (2011, 2014).

In this section, we compare the standard DEA efficiency scores of *provision* (Table 3.8) and *delivery* (Table 3.10) with the NDEA efficiency scores of Table 7.2 and reproduce these results in Table 7.3.

TABLE 7.2 Efficiency scores for node 1, 2 and centralized NDEA model, CRS input orientation.

```
RStudio

File   Edit   Code   View   Plots   Session   Build   Debug   Profile   Tools   Help

○ ▾  ○   ▾              Go to file/function          ▾ Addins ▾

Console   Terminal                                                              

C:/Program Files/RStudio/NDEA/

> data.frame(Eff_cent.node1= result.coop$eff.s1, Eff_cent.node2=
result.coop$eff.s2,Eff_cent_overall)
    Eff_cent.node1 Eff_cent.node2 Eff_cent_overall
1       0.8941651      0.9097854        0.8134983
2       0.5650345      0.6686039        0.3777843
3       0.5699745      0.9281796        0.5290387
4       0.7402270      0.7701017        0.5700501
5       0.6206234      0.9887013        0.6136112
6       0.8749553      0.8680382        0.7594946
7       1.0000000      0.8149268        0.8149268
8       0.9441422      1.0000000        0.9441422
9       0.3555824      0.9447873        0.3359497
10      0.7740209      0.9765008        0.7558320
11      0.5125106      1.0000000        0.5125106
12      0.7850431      0.7740137        0.6076340
13      1.0000000      1.0000000        1.0000000
14      0.5500696      0.7590640        0.4175380
15      0.2631544      0.7857945        0.2067852
16      1.0000000      0.9356125        0.9356125
>
```

TABLE 7.3 Efficiency scores for standard DEA efficiency scores and the centralized NDEA model, CRS input orientation.

	Provision	Delivery	Provision –delivery DEA	Node 1	Node 2	Centralized NDEA
	(1)	(2)	(3)	(4)	(5)	(6)
Air Canada	0.872	0.910	0.878	0.894	0.910	0.813
All Nippon Airways (ANA)	0.650	0.729	0.471	0.565	0.669	0.378
American Airlines	0.553	0.928	0.565	0.570	0.928	0.529
British Airways	0.624	0.910	0.735	0.740	0.770	0.570
Delta Air Lines	0.602	0.989	0.665	0.621	0.989	0.614
Emirates	0.875	0.978	0.966	0.875	0.868	0.759
Garuda Indonesia	0.623	0.933	0.912	1.000	0.815	0.815
KLM	0.944	1.000	1.000	0.944	1.000	0.944
Lufthansa	0.356	0.945	0.390	0.356	0.945	0.336
Malaysia Airlines	0.774	0.977	0.859	0.774	0.977	0.756
Qantas	0.482	1.000	0.621	0.513	1.000	0.513
SAS Scandinavian Airlines	0.590	0.969	0.649	0.785	0.774	0.608
Singapore Airlines	1.000	1.000	1.000	1.000	1.000	1.000
TAM	0.442	0.877	0.450	0.550	0.759	0.418
Thai Airways	0.225	0.882	0.239	0.263	0.786	0.207
United Airlines	1.000	0.936	1.000	1.000	0.936	0.936

In Table 7.3, column (3) efficiency scores (provision—delivery DEA) are based on R script of Table 3.3 using data set of Appendix E Table E.1, which considers only inputs x and outputs y and ignores the intermediates z. What we observe from Table 7.3 is that under standard DEA, there are more efficient airlines compared with the NDEA model of column 6. This suggests that when intermediates are unaccounted for in the production process, more airlines are observed to be efficient, which will not be the case when intermediates are accounted for. As the NDEA model is a product of node 1 and node 2 (see Section 7.3), then a DMU is only efficient if both nodes equal to 1. From Table 7.3, we observe that only Singapore Airlines is efficient (with nodes 1 and 2 equaling 1). When standard DEA is employed to derived efficiency scores for each activity (i.e. *provision* and *delivery*) independently, we observe that the *delivery* efficiency scores are generally higher than node 2, suggesting that the inefficiencies in node 1 of the NDEA model have an impact on the efficiency performance of node 2. This result indicates that the centralized NDEA model has more discriminate power than does the standard DEA model and is able to evaluate an airline's performance with respect to its maintenance and servicing, and fleet operation efficiency and performance on passenger and freight and mail tonne kilometres performed.

Chen et al. (2009, 2010) noted that a major limitation of Kao and Huang's (2008) model is that it is only applicable under CRS and its inability to handle the numerous situations under VRS. Also, its output-oriented efficiency score is not equal to the reciprocal of its input-oriented efficiency score for all DMUs. As noted in Fukuyama and Weber (2012), a shortcoming of the network models of Liang et al. (2008) and Seiford and Zhu (1999) are that outputs and inputs are scaled proportionally to the production frontier, as they are based on the Farrell efficiency radial model.[4]

7.4 Network DEA (Farrell efficiency model)—network technical efficiency

In this section, we introduce the input- and output-oriented network DEA Farrell efficiency model of Fukuyama and Matousek (2011) under VRS following the production model of Fig. 7.1. We term this as network technical efficiency (NTE).

---

4. The shortcoming also applies to Kao and Hwang (2008) network DEA model.

7.4.1 NTE input-oriented VRS model

$$
\text{NTE}_I = (x_o, y_o) = \min_{z, \lambda^1, \lambda^2, \theta} \left\{ \theta \left| \begin{array}{l} \theta x_o \geq \sum_{j=1}^{J} x_j \lambda_j^1; \ \sum_{j=1}^{J} x_j \lambda_j^1 \geq z; \ \sum_{j=1}^{J} x_j \lambda_j^2 \leq z; \\[3ex] y_o \leq \sum_{j=1}^{J} y_j \lambda_j^2; \ \sum_{j=1}^{J} \lambda_j^1 = 1; \ \sum_{j=1}^{J} \lambda_j^2 = 1; \\[3ex] z \geq 0; \ \lambda^1 \geq 0; \ \lambda^2 \geq 0; \ \theta: \textit{free} \end{array} \right. \right\}
$$

(7.6)

7.4.2 NTE output-oriented VRS model

$$
\text{NTE}_O = (x_o, y_o) = \max_{z, \lambda^1, \lambda^2, \varphi} \left\{ \varphi \left| \begin{array}{l} \sum_{j=1}^{J} z_j \lambda_j^1 \geq z; \ \sum_{j=1}^{J} z_j \lambda_j^2 \leq z; \ \phi y_o \leq \sum_{j=1}^{J} y_j \lambda_j^2; \\[3ex] x_o \geq \sum_{j=1}^{J} x_j \lambda_j^1; \ \sum_{j=1}^{J} \lambda_j^1 = 1; \ \sum_{j=1}^{J} \lambda_j^2 = 1; \\[3ex] z \geq 0; \ \lambda^1 \geq 0; \ \lambda^2 \geq 0; \ \theta: \text{free} \end{array} \right. \right\}
$$

(7.7)

The notations for the aforementioned equations are as follows. x_{io} ($i = 1$, ..., N) and y_{mo} ($m = 1, ..., M$) are the input and output quantities for DMU_o ($o = 1, ..., J$), respectively. For DMU$_j$, $x_j = (x_{1j}, ..., x_{Nj})$, where $j = 1, ..., J$ is an N-dimensional vector of inputs and $y_j = (y_{1j}, ..., y_{Mj})$, where $m = 1, ..., M$ is an M-dimensional vector of outputs. λ is an intensity variable of J-dimensional vector. z is a Q-dimensional vector of intermediate products, where $z = (z_1, ..., z_Q)^T \varepsilon \Re_+^Q$. I and O refer to input and output orientation, respectively. For CRS model, the notations $\sum_{j=1}^{J} \lambda_j^1 = 1$ and $\sum_{j=1}^{J} \lambda_j^2 = 1$ become $\sum_{j=1}^{J} \lambda_j^1 \geq 0$ and $\sum_{j=1}^{J} \lambda_j^2 \geq 0$.

7.4.3 R script for NTE input- and output-oriented VRS

The R script for NTE input-oriented VRS model is shown in Table 7.4 and the NTE output-oriented VRS model in Table 7.5.

TABLE 7.4 R script for NTE (input-oriented VRS).

```
## Load data (Reading from C drive and folder of data file)
data <- read.table(file="C:/Airline/NDEA1.txt", header=T,sep="\t")
attach(data)
X = cbind(x1,x2)
Y = cbind(y1,y2)
Z = cbind(z1,z2,z3)

nfi = function(xo,yo,Xref,Yref,Zref){

library(lpSolve)
n = nrow(Xref)
p = ncol(Xref)
q = ncol(Yref)
r = ncol(Zref)

obj.fun = c(1,rep(0,n+n+r))
constr.x = cbind(-xo,t(Xref),matrix(0,p,n+r))
constr.zx = cbind(matrix(0,r,1),-t(Zref),matrix(0,r,n),diag(r))
constr.zy = cbind(matrix(0,r,1+n),t(Zref),-diag(r))
constr.y = cbind(matrix(0,q,1+n),-t(Yref),matrix(0,q,r))
constr.lb = -diag(nrow=1+n+n+r)
constr.lambda1 = cbind(0,matrix(1,1,n),matrix(0,1,n+r))
constr.lambda2 = cbind(matrix(0,1,1+n),matrix(1,1,n),matrix(0,1,r))
constr = rbind(constr.x,constr.zx,constr.zy,constr.y,constr.lb,constr.lambda1,constr.lambda2)
constr.dir = c(rep("<=",p+r+r+q+1+n+n+r),"=","=")
rhs = c(rep(0,p+r+r),-yo,rep(0,1+n+n+r),1,1)

results = lp("min",obj.fun,constr,constr.dir,rhs)

return(results$objval)
}
nfi.all = function(X,Y,Xref,Yref,Zref){
m = nrow(X)
nfi = rep(0,m)
for (i in 1:m) {nfi[i] = nfi(X[i,],Y[i,],Xref,Yref,Zref)}
return(round(nfi,5))
}
eff_score<- nfi.all(X,Y,X,Y,Z)
data.frame(dmu, eff_score)
```

> "=","=" refers to VRS. As noted in the previous line, constraint lambda1 and constraint lambda 2, these are equal to 1 in the next line (1,1).
>
> To change to CRS, both signs must be >=, and the values changed to 0 for both.

7.4.4 Results for the NTE input- and output-oriented VRS and CRS model

The results for the NTE input- and output-oriented VRS and CRS models are presented from Tables 7.6, 7.7, 7.8 and 7.9. Note that while only the R script for VRS model is provided, changing to CRS model is a straight-forward process and explained in Tables 7.4 and 7.5.

Comparing the efficiency scores between the centralized NDEA model of Table 7.2 and the NTE input-oriented CRS model of Table 7.8, we note that

TABLE 7.5 R script for NTE (output-oriented VRS).

```
## Load data (Reading from C drive and folder of data file)
data <- read.table(file="C:/Airline/NDEA1.txt", header=T,sep="\t")
attach(data)
X = cbind(x1,x2)
Y = cbind(y1,y2)
Z = cbind(z1,z2,z3)

nfo = function(xo,yo,Xref,Yref,Zref){

  library(lpSolve)
  n = nrow(Xref)
  p = ncol(Xref)
  q = ncol(Yref)
  r = ncol(Zref)

  obj.fun = c(1,rep(0,n+n+r))
  constr.x = cbind(matrix(0,nrow=p,ncol=1),t(Xref),matrix(0,nrow=p,ncol=n+r))
  constr.zx = cbind(matrix(0,nrow=r,ncol=1),-t(Zref),matrix(0,nrow=r,ncol=n),diag(nrow=r))
  constr.zy = cbind(matrix(0,nrow=r,ncol=1+n),t(Zref),-diag(nrow=r))
  constr.y = cbind(yo,matrix(0,nrow=q,ncol=n),-t(Yref),matrix(0,nrow=q,ncol=r))
  constr.lb = -diag(nrow=1+n+n+r)
  constr.lambda1 = cbind(0,matrix(1,nrow=1,ncol=n),matrix(0,nrow=1,ncol=n+r))
  constr.lambda2 =
cbind(matrix(0,nrow=1,ncol=1+n),matrix(1,nrow=1,ncol=n),matrix(0,nrow=1,ncol=r))
  constr = rbind(constr.x,constr.zx,constr.zy,constr.y,constr.lb,constr.lambda1,constr.lambda2)
  constr.dir = c(rep("<=",p+r+r+q+1+n+n+r),"=","=")
  rhs = c(xo,rep(0,r+r+q+1+n+n+r),1,1)

  results = lp("max",obj.fun,constr,constr.dir,rhs)
  return(results$objval)

}
nfo.all = function(X,Y,Xref,Yref,Zref){
  m = nrow(X)
  nfo = rep(0,m)
  for (i in 1:m) {nfo[i] = nfo(X[i,],Y[i,],Xref,Yref,Zref)}
  return(round(nfo,5))
}
eff_score<- nfo.all(X,Y,X,Y,Z)
data.frame(dmu, eff_score)
```

> "=","=" refers to VRS. As noted in the previous line, constraint lambda1 and constraint lambda 2, these are equal to 1 in the next line (1,1).
>
> To change to CRS, both signs must be >=, and the values changed to 0 for both.

the results are identical. We observe that under CRS, the input-oriented efficiency scores are equal to the reciprocal of the output-oriented efficiency scores. However, under VRS, this is not the case, which suggests that the NTE model faces similar problems to the Kao and Hwang (2008) and Liang et al. (2008) models. Furthermore, as NTE is an extension of the Farrell efficiency model, which is a radial model, the assumption that inputs or outputs undergo proportional changes may not be true. Hence, while NTE observes the ratio efficiency, it does not account for the slacks.

TABLE 7.6 Efficiency scores for NTE VRS input orientation.

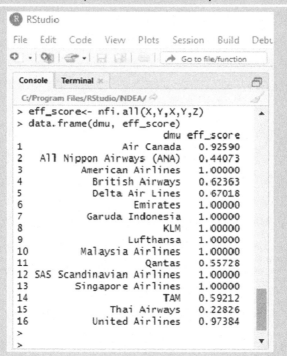

```
R RStudio

File   Edit   Code   View   Plots   Session   Build   Debu

  ⊙  ·  ⊠  ⌂ ·  ⊟  ⊞  ⊟  |  ⇨ Go to file/function

 Console   Terminal ×                                        ⊟

 C:/Program Files/RStudio/NDEA/

> eff_score<- nfi.all(X,Y,X,Y,Z)
> data.frame(dmu, eff_score)
                          dmu eff_score
1                   Air Canada   0.92590
2       All Nippon Airways (ANA)   0.44073
3             American Airlines   1.00000
4               British Airways   0.62363
5               Delta Air Lines   0.67018
6                      Emirates   1.00000
7              Garuda Indonesia   1.00000
8                           KLM   1.00000
9                     Lufthansa   1.00000
10            Malaysia Airlines   1.00000
11                       Qantas   0.55728
12   SAS Scandinavian Airlines   1.00000
13            Singapore Airlines   1.00000
14                          TAM   0.59212
15                  Thai Airways   0.22826
16              United Airlines   0.97384
>
>
```

TABLE 7.7 Efficiency scores for NTE VRS output orientation.

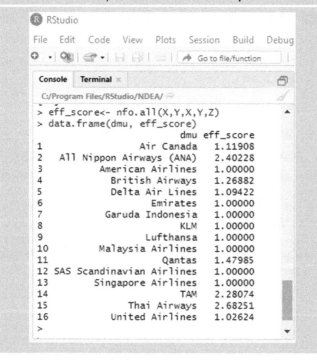

```
R RStudio

File   Edit   Code   View   Plots   Session   Build   Debug

  ⊙  ·  ⊠  ⌂ ·  ⊟  ⊞  ⊟  |  ⇨ Go to file/function

 Console   Terminal ×                                        ⊟

 C:/Program Files/RStudio/NDEA/

> eff_score<- nfo.all(X,Y,X,Y,Z)
> data.frame(dmu, eff_score)
                          dmu eff_score
1                   Air Canada   1.11908
2       All Nippon Airways (ANA)   2.40228
3             American Airlines   1.00000
4               British Airways   1.26882
5               Delta Air Lines   1.09422
6                      Emirates   1.00000
7              Garuda Indonesia   1.00000
8                           KLM   1.00000
9                     Lufthansa   1.00000
10            Malaysia Airlines   1.00000
11                       Qantas   1.47985
12   SAS Scandinavian Airlines   1.00000
13            Singapore Airlines   1.00000
14                          TAM   2.28074
15                  Thai Airways   2.68251
16              United Airlines   1.02624
>
```

TABLE 7.8 Efficiency scores for NTE CRS input orientation.

```
R RStudio

File   Edit   Code   View   Plots   Session   Build   Deb

○ · ○ · ⊟ ⊡ ⊟       ➔ Go to file/function

Console   Terminal ×

C:/Program Files/RStudio/NDEA/

> eff_score<- nfi.all(X,Y,X,Y,Z)
> data.frame(dmu, eff_score)
                              dmu eff_score
1                      Air Canada   0.81350
2        All Nippon Airways (ANA)   0.37778
3               American Airlines   0.52904
4                 British Airways   0.57005
5                Delta Air Lines   0.61361
6                        Emirates   0.75949
7                Garuda Indonesia   0.81493
8                            KLM   0.94414
9                      Lufthansa   0.33595
10              Malaysia Airlines   0.75583
11                        Qantas   0.51251
12   SAS Scandinavian Airlines    0.60763
13             Singapore Airlines   1.00000
14                            TAM   0.41754
15                  Thai Airways   0.20679
16                United Airlines   0.93561
>
```

7.5 Network cost efficiency model (Fukuyama and Matousek, 2011)

Following the production model of Fig. 7.1, the network cost efficiency (NCE) function of Fukuyama and Matousek (2011) is expressed as

$$
\text{NCE}_o = \min_{x, z, \lambda^1, \lambda^2} \left\{ \frac{c\widehat{x}^*}{\cos t_o} \left| \begin{array}{l} x \geq \sum_{j=1}^{J} x_j \lambda_j^1; \quad \sum_{j=1}^{J} z \lambda_j^1 \geq z; \quad \sum_{j=1}^{J} z_j \lambda_j^2 \leq z; \\[2mm] y_o \leq \sum_{j=1}^{J} y_j \lambda_j^2; \quad \sum_{j=1}^{J} \lambda_j^1 = 1; \quad \sum_{j=1}^{J} \lambda_j^2 = 1; \\[2mm] z \geq 0; \quad x \geq 0; \quad \lambda^1 \geq 0; \quad \lambda^2 \geq 0 \end{array} \right. \right\} \tag{7.8}
$$

where $\widehat{x}$ indicates a vector of variables, c_n ($n = 1, \dots, N$) is the price of input n and the inner product, and $cx_o = c_1 x_{1o} + \dots + c_N x_{No} = \cos t_o$ is the observed total cost for DMU_o. Hence, $\text{NCE}_o = c\widehat{x}^*$ is the optimal objective. All other notations follow those described in Section 7.4.

TABLE 7.9 Efficiency scores for NTE CRS output orientation.

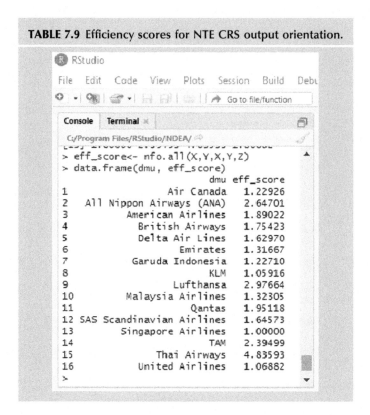

```
> eff_score<- nfo.all(X,Y,X,Y,Z)
> data.frame(dmu, eff_score)
                               dmu eff_score
1                        Air Canada   1.22926
2       All Nippon Airways (ANA)     2.64701
3                 American Airlines   1.89022
4                   British Airways   1.75423
5                  Delta Air Lines    1.62970
6                          Emirates   1.31667
7                 Garuda Indonesia    1.22710
8                               KLM   1.05916
9                        Lufthansa    2.97664
10                Malaysia Airlines   1.32305
11                          Qantas    1.95118
12      SAS Scandinavian Airlines    1.64573
13                Singapore Airlines   1.00000
14                              TAM   2.39499
15                    Thai Airways    4.83593
16                  United Airlines   1.06882
>
```

7.5.1 R script for NCE VRS model

The R script for NCE VRS model is shown in Table 7.10.

As input prices were unavailable and the inconsistent definitions across airline annual reports meant that data from such sources would result in unreliable unit prices, we therefore used PPPs as a proxy for input prices.

7.5.2 Results for the NCE VRS model

The results for the NCE VRS model is based on the data set of Appendix E Table E.2.[5]

From Table 7.11, optCost refers to the numerator of 'nce' in Table 7.10, which is the optimal objective, $c\widehat{x}^*$ in Eq. (7.8). This means that if $c\widehat{x}^*$ equals to the observed total cost, cos t_o, then the value of NCE equals to 1. This is the

---

5. As cost efficiency is based on the product of input quantities and input prices, it is important to ensure that the units in the data set are in full otherwise aggregation without factoring in columns showing '000 will lead to inaccurate optimal targets.

TABLE 7.10 R script for NCE (VRS).

```
## Load data (Reading from C drive and folder of data file)
data <- read.table(file="C:/Airline/NDEA2.txt ", header=T,sep="\t")
attach(data)
X = cbind(x1,x2)
Y = cbind(y1,y2)
Z = cbind(z1,z2,z3)
W = cbind(w1,w2)

nce = function(xo,yo,wo,Xref,Yref,Zref){

  library(lpSolve)
  n = nrow(Xref)
  p = ncol(Xref)
  q = ncol(Yref)
  r = ncol(Zref)

  obj.fun = c(wo,rep(0,n+n+r))
  constr.x = cbind(-diag(p),t(Xref),matrix(0,p,n+r))
  constr.zx = cbind(matrix(0,r,p),-t(Zref),matrix(0,r,n),diag(r))
  constr.zy = cbind(matrix(0,r,p+n),t(Zref),-diag(r))
  constr.y = cbind(matrix(0,q,p+n),-t(Yref),matrix(0,q,r))
  constr.lb = -diag(p+n+n+r)
  constr.lambda1 = cbind(matrix(0,1,p),matrix(1,1,n),matrix(0,1,n+r))
  constr.lambda2 = cbind(matrix(0,1,p+n),matrix(1,1,n),matrix(0,1,r))
  constr = rbind(constr.x,constr.zx,constr.zy,constr.y,constr.lb,constr.lambda1,constr.lambda2)
  constr.dir = c(rep("<=",p+r+r+q+p+n+n+r),"=","=")
  rhs = c(rep(0,p+r+r),-yo,rep(0,p+n+n+r),1,1)

  results = lp("min",obj.fun,constr,constr.dir,rhs)

  optCost = results$objval
  optX = results$solution[1:p]
  nce = optCost/sum(wo*xo)

  return(list(nce=nce,optCost=optCost,optX=optX))

}
nce.all = function(X,Y,W,Xref,Yref,Zref){
  m = nrow(X)
  p = ncol(X)
  results = list(nce=rep(0,m),optCost=rep(0,m),optX=matrix(0,m,p))
  for (i in 1:m){
    temp = nce(X[i,],Y[i,],W[i,],Xref,Yref,Zref)
    results$nce[i] = round(temp$nce,5)
    results$optCost[i] = round(temp$optCost,5)
    results$optX[i,] = round(temp$optX,5)
  }
  return(results)
}
results<- nce.all(X,Y,W,X,Y,Z)
data.frame(dmu, results)
```

TABLE 7.11 Results of NCE VRS model.

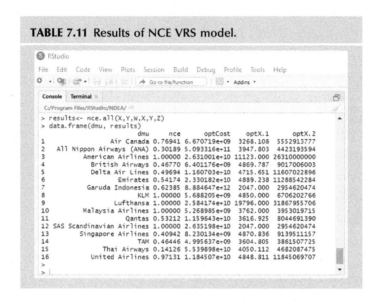

```
> results<- nce.all(X,Y,W,X,Y,Z)
> data.frame(dmu, results)
                             dmu     nce       optCost     optX.1        optX.2
1                      Air Canada 0.76941 6.670719e+09  3268.108    5552913777
2            All Nippon Airways (ANA) 0.30189 5.093316e+11  3947.803    4423193594
3                American Airlines 1.00000 2.631001e+10 11123.000   26310000000
4                   British Airways 0.46770 6.401176e+09  4869.787    9017006003
5                   Delta Air Lines 0.49694 1.160703e+10  4715.651   11607022896
6                          Emirates 0.54174 2.330182e+10  4889.238   11288542284
7                  Garuda Indonesia 0.62385 8.884647e+12  2047.000    2954620474
8                               KLM 1.00000 5.688205e+09  4850.000    6706202766
9                         Lufthansa 1.00000 2.584174e+10 19796.000   31867955706
10                 Malaysia Airlines 1.00000 5.268985e+09  3762.000    3953019715
11                            Qantas 0.53212 1.159643e+10  3616.925    8044691390
12       SAS Scandinavian Airlines 1.00000 2.635198e+10  2047.000    2954620474
13                 Singapore Airlines 0.40942 8.230134e+09  4870.836    9139511157
14                               TAM 0.46446 4.995637e+09  3604.805    3861507725
15                      Thai Airways 0.14126 5.539898e+10  4050.112    4682087475
16                   United Airlines 0.97131 1.184507e+10  4848.811   11845069707
>
> |
```

case for American Airlines, KLM, Lufthansa, Malaysia Airlines and SAS. 'optCost' shows the target cost airlines should achieve to become efficient. For Air Canada, the 'optCost' of 6.670719e+09 (or US\$ 6,670,719,000) suggests that the airline should increase its $x1$ from 2755 to 3268 and decrease its $x2$ from 8,669,927,829 to 5,552,913,777. The targeted total cost would be $x1w1 + x2w2 = (3268 \times 1.2013) + (5,552,913,777 \times 1.2013) = \$US$ 6,670,719,246.

Fukuyama and Matousek (2011) showed that the NCE model can be decomposed into the product of network allocative efficiency (NAE) and network technical efficiency (NTE):

$$NCE = NAE \times NTE \tag{7.9}$$

whereby NAE is derived residually from Eq. (7.9) to

$$NCE = NAE/NTE \tag{7.10}$$

The NCE, NAE and NTE for each airline are shown in Table 7.12.

Fukuyama and Matousek (2011) showed that the cost efficiency bias is derived by dividing Eq. (7.9) with Eq. (7.12):

$$CE = AE \times TE \tag{7.12}$$

which is derived by rearranging Eq. (3.5) of Chapter 3.

The following expression is then obtained:

TABLE 7.12 NCE, NAE and NTE (VRS model).

	NCE	NAE	NTE
Air Canada	0.7694	0.8310	0.9259
All Nippon Airways (ANA)	0.3019	0.6850	0.4407
American Airlines	1.0000	1.0000	1.0000
British Airways	0.4677	0.7500	0.6236
Delta Air Lines	0.4969	0.7415	0.6702
Emirates	0.5417	0.5417	1.0000
Garuda Indonesia	0.6239	0.6239	1.0000
KLM	1.0000	1.0000	1.0000
Lufthansa	1.0000	1.0000	1.0000
Malaysia Airlines	1.0000	1.0000	1.0000
Qantas	0.5321	0.9549	0.5573
SAS Scandinavian Airlines	1.0000	1.0000	1.0000
Singapore Airlines	0.4094	0.4094	1.0000
TAM	0.4645	0.7844	0.5921
Thai Airways	0.1413	0.6189	0.2283
United Airlines	0.9713	0.9974	0.9738

Source: NCE from Table 7.11 and NTE from Table 7.6.

$$\frac{\text{NCE}}{\text{CE}} = \frac{\text{NAE}}{\text{AE}} \times \frac{\text{NTE}}{\text{TE}} \qquad (7.13)$$

whereby $\frac{\text{NCE}}{\text{CE}}$ is the cost efficiency (CE) bias, $\frac{\text{NAE}}{\text{AE}}$ is the allocative efficiency (AE) bias and $\frac{\text{NTE}}{\text{TE}}$ is the technical efficiency (TE) bias. Decomposition of Eq. (7.13) into bias estimates allows for identification of causes of cost efficiency measurement bias.

To illustrate the CE bias, we first estimate the standard CE, AE and TE under VRS using the data set of Appendix E Table E.2. Using R script in Table 3.9, and considering only inputs x and outputs y in the production model, we obtain the following results.

Following Eq. (7.12), we then have the results of CE bias, AE bias and TE bias as shown in Table 7.14.

Using Air Canada as an example, NCE = 0.76962 from Table 7.12 divided by CE = 0.8340457 from Table 7.13 equals 0.9228. NTE = 0.9259 from Table 7.12 divided by TE = 0.985928 from Table 7.13 equals 0.9391. NAE/

TABLE 7.13 Results of cost efficiency, allocative efficiency and technical efficiency VRS model (excluding intermediates).

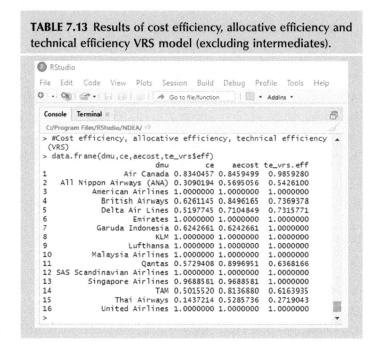

```
R RStudio

File   Edit   Code   View   Plots   Session   Build   Debug   Profile   Tools   Help

  ⚙ ▾ ⚙▾ ⚙ ▾    📄 📄   📄    ↗ Go to file/function        ▦ ▾ Addins ▾

Console   Terminal ⊗                                                          🗗

C:/Program Files/RStudio/NDEA/ ⇗
> #Cost efficiency, allocative efficiency, technical efficiency ▲
(VRS)
> data.frame(dmu,ce,aecost,te_vrs$eff)
                         dmu        ce    aecost te_vrs.eff
1                  Air Canada 0.8340457 0.8459499  0.9859280
2      All Nippon Airways (ANA) 0.3090194 0.5695056  0.5426100
3            American Airlines 1.0000000 1.0000000  1.0000000
4              British Airways 0.6261145 0.8496165  0.7369378
5              Delta Air Lines 0.5197745 0.7104849  0.7315771
6                     Emirates 1.0000000 1.0000000  1.0000000
7             Garuda Indonesia 0.6242661 0.6242661  1.0000000
8                          KLM 1.0000000 1.0000000  1.0000000
9                   Lufthansa 1.0000000 1.0000000  1.0000000
10           Malaysia Airlines 1.0000000 1.0000000  1.0000000
11                     Qantas 0.5729408 0.8996951  0.6368166
12 SAS Scandinavian Airlines 1.0000000 1.0000000  1.0000000
13           Singapore Airlines 0.9688581 0.9688581  1.0000000
14                         TAM 0.5015520 0.8136880  0.6163935
15                 Thai Airways 0.1437214 0.5285736  0.2719043
16             United Airlines 1.0000000 1.0000000  1.0000000
>
```

$AE = (NCE/NTE)/(CE/TE) = (0.76962/0.9259)/(0.8340457/0.985928) = 0.9826$.

7.6 Network revenue efficiency model (Fukuyama and Matousek, 2017)

The following efficiency model, network revenue efficiency (NRE), follows the network revenue function of Fukuyama and Matousek (2017), excluding bad output. Following the production model of Fig. 7.1 and notations of Section 7.4, the NRE is expressed as follows:

$$\text{NRE}_o = \max_{y,z,\lambda^1,\lambda^2} \left\{ p^T y \left| \begin{array}{l} x_o \geq \sum_{j=1}^{J} x_j \lambda_j^1; \quad \sum_{j=1}^{J} z\lambda_j^1 \geq z; \quad \sum_{j=1}^{J} z_j \lambda_j^2 \leq z; \\[2ex] y \leq \sum_{j=1}^{J} y_j \lambda_j^2; \quad \sum_{j=1}^{J} \lambda_j^1 = 1; \quad \sum_{j=1}^{J} \lambda_j^2 = 1; \\[2ex] z \geq 0; \quad x \geq 0; \quad \lambda^1 \geq 0; \quad \lambda^2 \geq 0 \end{array} \right. \right\} \quad (7.14)$$

where $p^T y$ is total revenue $= p_1 y_1 + \dots p_m y_m + \dots + p_M y_M$.

TABLE 7.14 Results of CE bias, AE bias and TE bias (VRS model).

	NCE/CE	NAE/AE	NTE/TE
	(CE bias)	(AE bias)	(TE bias)
Air Canada	0.9228	0.9826	0.9391
All Nippon Airways (ANA)	0.9776	1.2036	0.8122
American Airlines	1.0000	1.0000	1.0000
British Airways	0.7472	0.8830	0.8462
Delta Air Lines	0.9562	1.0438	0.9161
Emirates	0.5419	0.5419	1.0000
Garuda Indonesia	1.0000	1.0000	1.0000
KLM	1.0000	1.0000	1.0000
Lufthansa	1.0000	1.0000	1.0000
Malaysia Airlines	1.0000	1.0000	1.0000
Qantas	0.9288	1.0614	0.8751
SAS Scandinavian Airlines	1.0000	1.0000	1.0000
Singapore Airlines	0.4228	0.4228	1.0000
TAM	0.9266	0.9646	0.9606
Thai Airways	0.9836	1.1716	0.8395
United Airlines	0.9713	0.9974	0.9738

7.6.1 R script for NRE VRS model

The R script for NCE VRS model is shown in Table 7.15.

7.6.2 Results for the NRE VRS model

The results for the NCE VRS model are based on the data set of Appendix E Table E.3.[6]

In similar fashion to the previous section, Table 7.16 provides information of optimal targets relating to revenue efficiency. For Air Canada, the 'optRev' of US$9,163,183,616 suggests that the airline should increase its $y1$ from 6,420,786,000 tonne km to 7,185,395,988 tonne km and increase its $y2$ from 1,157,081,000 tonne km to 2,267,405,205 tonne km. The targeted total

---

6. As revenue efficiency is based on the product of output quantities and output prices, it is important to ensure that the units in the data set are in full otherwise aggregation without factoring in columns showing '000 will lead to inaccurate optimal targets.

TABLE 7.15 R script for NRE (VRS).

```
## Load data (Reading from C drive and folder of data file)
data <- read.table(file="C:/Airline/NDEA3.txt ", header=T,sep="\t")
attach(data)
X = cbind(x1,x2)
Y = cbind(y1,y2)
Z = cbind(z1,z2,z3)
V = cbind(v1,v2)

nre = function(xo,yo,vo,Xref,Yref,Zref){

 library(lpSolve)
 n = nrow(Xref)
 p = ncol(Xref)
 q = ncol(Yref)
 r = ncol(Zref)

 obj.fun = c(vo,rep(0,n+n+r))
 constr.x = cbind(matrix(0,p,q),t(Xref),matrix(0,p,n+r))
 constr.zx = cbind(matrix(0,r,q),-t(Zref),matrix(0,r,n),diag(r))
 constr.zy = cbind(matrix(0,r,q+n),t(Zref),-diag(r))
 constr.y = cbind(diag(q),matrix(0,q,n),-t(Yref),matrix(0,q,r))
 constr.lb = -diag(q+n+n+r)
 constr.lambda1 = cbind(matrix(0,1,q),matrix(1,1,n),matrix(0,1,n+r))
 constr.lambda2 = cbind(matrix(0,1,q+n),matrix(1,1,n),matrix(0,1,r))
 constr = rbind(constr.x,constr.zx,constr.zy,constr.y,constr.lb,constr.lambda1,constr.lambda2)
 constr.dir = c(rep("<=",p+r+r+q+q+n+n+r),"=","=")
 rhs = c(xo,rep(0,r+r+q+q+n+n+r),1,1)

 results = lp("max",obj.fun,constr,constr.dir,rhs)

 optRev = results$objval
 optY = results$solution[1:q]
 nre = sum(vo*yo)/optRev

 return(list(nre=nre,optRev=optRev,optY=optY))

}
nre.all = function(X,Y,V,Xref,Yref,Zref){

 m = nrow(X)
 q = ncol(Y)
 results = list(nre=rep(0,m),optRev=rep(0,m),optY=matrix(0,m,q))
 for (i in 1:m){
   temp = nre(X[i,],Y[i,],V[i,],Xref,Yref,Zref)
   results$nre[i] = round(temp$nre,5)
   results$optRev[i] = round(temp$optRev,5)
   results$optY[i,] = round(temp$optY,5)
 }
 return(results)
}
results<- nre.all(X,Y,V,X,Y,Z)
data.frame(dmu, results)
```

TABLE 7.16 Results of NRE VRS model.

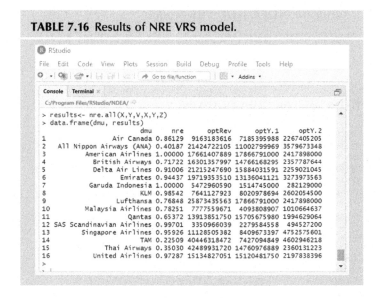

```
> results<- nre.all(X,Y,V,X,Y,Z)
> data.frame(dmu, results)
                             dmu     nre     optRev      optY.1      optY.2
1                      Air Canada 0.86129 9163183616  7185395988  2267405205
2       All Nippon Airways (ANA) 0.40187 21424722105 11002799969 3579673348
3              American Airlines 1.00000 17661407889 17866791000 2417898000
4                British Airways 0.71722 16301357997 14766168295 2357787644
5                Delta Air Lines 0.91006 21215247690 15884031591 2259021045
6                       Emirates 0.94437 19719353510 13136041121 3273973563
7               Garuda Indonesia 1.00000 5472960590  1514745000  282129000
8                            KLM 0.98542 7641127923  8020978694  2602054500
9                     Lufthansa 0.76848 25873435563 17866791000 2417898000
10             Malaysia Airlines 0.78251 7777559671  4093808907  1010664637
11                        Qantas 0.65372 13913851750 15705675980 1994629064
12 SAS Scandinavian Airlines 0.99701 3350966039  2279584558  494527200
13             Singapore Airlines 0.95926 11128505382 8409673397  4752575601
14                           TAM 0.22509 40446318472 7427094849  4602946218
15                  Thai Airways 0.35030 42489931720 14760976889 2360131223
16                United Airlines 0.97287 15134827051 15120481750 2197838396
>
```

revenue would be $y1v1 + y2v2 = (7,185,395,988 \times 1.167781027) + (2,267,405,205 \times 0.340571905) = \$US9,163,183,617.$

7.7 Network DEA directional distance function inefficiency model (Fukuyama and Weber, 2012)

The following network DEA directional distance function (NDEA-DDF) inefficiency model is based on Fukuyama and Weber (2012) nonoriented network DDF. Following the production model of Fig. 7.1 and notations of Section 7.4, the NDEA-DDF is expressed as follows:

$$
\overrightarrow{\mathrm{ND}}(x_o, y_o; g^x, g^y) = \max_{z, \lambda^1, \lambda^2, \delta} \left\{ \delta \left| \begin{array}{l} x_o - \delta g^x \geq \sum_{j=1}^{J} x_j \lambda_j^1; \quad \sum_{j=1}^{J} z\lambda_j^1 \geq z; \quad \sum_{j=1}^{J} z_j \lambda_j^2 \leq z; \\[2ex] y_o + \delta g^y \leq \sum_{j=1}^{J} y_j \lambda_j^2; \quad \sum_{j=1}^{J} \lambda_j^1 = 1; \quad \sum_{j=1}^{J} \lambda_j^2 = 1; \\[2ex] \quad\quad\quad z \geq 0; \quad \lambda^1 \geq 0; \quad \lambda^2 \geq 0; \quad \delta \text{ free} \end{array} \right. \right\}
$$

$$(7.15)$$

whereby the directional distance function expands outputs along the directional vector $g^y = (g_1^y, ..., g_M^y) = \in \Re_+^M$ and contracts inputs along the directional vector $g^x = (g_1^x, ..., g_N^x) = \in \Re_+^N$ and assuming the directional vectors $g^x = (g_1^x, g_2^x) = (1, 1)$ and $g^y = (g_1^y, g_2^y) = (1, 1)$.

7.7.1 R script for NDEA-DDF VRS model

The R script for NDEA-DDF VRS model is shown in Table 7.17.

TABLE 7.17 R script for NDEA-DDF (VRS).

```
## Load data (Reading from C drive and folder of data file)
data <- read.table(file="C:/Airline/NDEA1.txt", header=T,sep="\t")
attach(data)
X = cbind(x1,x2)
Y = cbind(y1,y2)
Z = cbind(z1,z2,z3)
gx = c(1,1)
gy = c(1,1)

nd = function(xo,yo,Xref,Yref,Zref,gx,gy){

  library(lpSolve)
  n = nrow(Xref)
  p = ncol(Xref)
  q = ncol(Yref)
  r = ncol(Zref)

  obj.fun = c(1,rep(0,n+n+r))
  constr.x = cbind(gx,t(Xref),matrix(0,nrow=p,ncol=n+r))
  constr.zx = cbind(matrix(0,nrow=r,ncol=1),-t(Zref),matrix(0,nrow=r,ncol=n),diag(nrow=r))
  constr.zy = cbind(matrix(0,nrow=r,ncol=1+n),t(Zref),-diag(nrow=r))
  constr.y = cbind(gy,matrix(0,nrow=q,ncol=n),-t(Yref),matrix(0,nrow=q,ncol=r))
  constr.lb = -diag(nrow=1+n+n+r)
  constr.lambda1 = cbind(0,matrix(1,nrow=1,ncol=n),matrix(0,nrow=1,ncol=n+r))
  constr.lambda2 =
cbind(matrix(0,nrow=1,ncol=1+n),matrix(1,nrow=1,ncol=n),matrix(0,nrow=1,ncol=r))
  constr = rbind(constr.x,constr.zx,constr.zy,constr.y,constr.lb,constr.lambda1,constr.lambda2)
  constr.dir = c(rep("<=",p+r+r+q+1+n+n+r),"=","=")
  rhs = c(xo,rep(0,r+r),-yo,rep(0,1+n+n+r),1,1)

  results = lp("max",obj.fun,constr,constr.dir,rhs)
  nd = results$objval
  lambda1 = results$solution[2:(n+1)]
  lambda2 = results$solution[(2+n):(1+n+n)]
  zopt = results$solution[(2+n+n):(1+n+n+r)]
  return(list(nd=nd,lambda1=lambda1,lambda2=lambda2,zopt=zopt))
  }
nd.all = function(X,Y,Xref,Yref,Zref,gx,gy){
  m = nrow(X)
  n = nrow(Xref)
  r = ncol(Zref)
  results = list(nd=rep(0,m), zopt=matrix(0,m,r))
  for (i in 1:m) {
    temp = nd(X[i,],Y[i,],Xref,Yref,Zref,gx,gy)
    results$nd[i] = round(temp$nd,5)
    results$zopt[i,] = temp$zopt
  }
  return(results)
}
results<- nd.all(X,Y,X,Y,Z,gx,gy)
data.frame(dmu, results)
```

7.7.2 Results for the NDEA-DDF VRS model

The results for the NDEA-DDF VRS model are presented in Table 7.18 and based on the data set of Appendix E Table E.1. Note that the ND is the inefficiency score of the network technology. If $ND = 0$, then the network technology for DMU_o is efficient; otherwise, it is inefficient.

Table 7.18 results show the same 8 airlines exhibiting efficiency as those in Table 7.6. In terms of inefficiency, we observe that the further the score is from 0, the more inefficient the airline is. Thai Airways is noted as the most inefficient among the sample airlines in both Tables 7.6 and 7.18. The zopt values provide the optimal values for $z1$, $z2$ and $z3$ intermediates for the inefficient airlines to target to raise their efficiency levels.

Fukuyama and Weber (2012) showed that network inefficiency (ND) can be decomposed into the sum of network bias (Net bias) and the black box directional inefficiency (DDF). The network bias indicator shows the degree of bias from estimating a black box directional distance function when a network structure exists. The results of the decomposition under VRS and CRS are shown in Table 7.19.

If the net bias indicator is positive, while $DDF = 0$, which exhibits efficiency such as Delta Airlines and Qantas, then these airline operators are liable of incorrectly exaggerating the true efficiency performance and lose out on the opportunity to enhance overall performance by accounting for the synergies between the two nodes '*provision*' and '*delivery*' within the firm.

TABLE 7.18 Results of NDEA-DDF VRS model.

```
RStudio

File  Edit  Code  View  Plots  Session  Build  Debug  Profile  Tools  Help

⊙ ▾ 🔍 ▾ 🖅 ▾ | 🔚 🔁 | 🔁 | 🗦 | ⋀ Go to file/function     | 🔡 ▾ Addins ▾

Console   Terminal ×

C:/Program Files/RStudio/NDEA/
> results<- nd.all(X,Y,X,Y,Z,gx,gy)
> data.frame(dmu, results)
                          dmu        nd       zopt.1      zopt.2     zopt.3
1                   Air Canada  505.0963  11782617239   7768.928  2261118.2
2           All Nippon Airways (ANA) 2358.0600   8642826205   5024.034  1782058.0
3             American Airlines    0.0000  34707729000  23102.000  8654892.9
4               British Airways 3041.5233  22850422792  11615.397  4293872.0
5               Delta Air Lines 2186.0491  27991888554  17542.248  7225936.0
6                     Emirates    0.0000  27369447000  13153.000  4717271.6
7             Garuda Indonesia    0.0000   2834184000   2187.000   676346.5
8                          KLM    0.0000  15090771000   8101.000  3027818.2
9                   Lufthansa    0.0000  27007957000  33288.000  5759785.6
10            Malaysia Airlines    0.0000   7292543000   5231.000  1606904.0
11                      Qantas 4022.7635  20653460547  11446.673  4444931.5
12    SAS Scandinavian Airlines    0.0000    415267000   4046.000  1108178.3
13           Singapore Airlines    0.0000  21286125000   9467.000  3513669.0
14                          TAM 2274.9488   7922325164   4642.110  1652525.3
15                 Thai Airways 4366.5910   9526416966   5467.528  1954906.5
16              United Airlines  219.0081  28199172969  17839.318  7110316.1
>
```

TABLE 7.19 DDF objective value, ND objective value and net bias (VRS and CRS).

	VRS				CRS		
	DDF	ND	Net bias		DDF	ND	Net bias
Air Canada	1138	505	−634		1210	650	−560
ANA All Nippon Airways	1482	2358	876		1753	2372	619
American Airlines	0	0	0		2627	7576	4949
British Airways	4827	3042	−1788		5449	3259	−2190
Delta Air Lines	0	2186	2186		316	4022	3706
Emirates	0	0	0		616	1778	1162
Garuda Indonesia	0	0	0		409	25	−384
KLM	0	0	0		0	2396	2396
Lufthansa	0	0	0		19,608	19,099	−509
Malaysia Airlines	0	0	0		1758	2789	1031
Qantas	0	4023	4023		0	4178	4178
SAS Scandinavian Airlines	0	0	0		1279	1517	238
Singapore Airlines	0	0	0		0	0	0
TAM	2130	2275	144		2374	2358	−17
Thai Airways	1931	4367	2435		2163	4378	2215
United Airlines	756	219	−537		1916	445	−1471

Source: DDF is derived using the R script for DDF in Chapter 5 Table 5.15 and data set of Appendix C.

7.8 Network slacks-based inefficiency model

The NSBI model of Fukuyama and Weber (2010), which accounts for bad outputs, is adapted from Tone and Tsutsui's (2009) network slacks-based measure. excluding the bad output component. Fukuyama and Weber (2010) defines the NSBI VRS model for DMU$_o$ as

$$
\text{NSBI}(x_o, y_o; g^x, g^y) = \max_{z, \lambda^1, \lambda^2, s^x, s^y} \left\{ \left. \frac{\frac{1}{N}\sum_{n=1}^{N}\frac{s_n^x}{g_n^x} + \frac{1}{M}\sum_{m=1}^{M}\frac{s_m^y}{g_m^y}}{2} \right| \begin{array}{c} x_o - s^x \geq \sum_{j=1}^{J}x_j\lambda_j^1; \quad \sum_{j=1}^{J}z_j\lambda_j^1 \geq z; \\[2mm] \sum_{j=1}^{J}z_j\lambda_j^2 \leq z; \quad y_o + s^y \leq \sum_{j=1}^{J}y_j\lambda_j^2; \\[2mm] \sum_{j=1}^{J}\lambda_j^1 = 1; \quad \sum_{j=1}^{J}\lambda_j^2 = 1; \\[2mm] z \geq 0; \quad \lambda^1 \geq 0; \quad \lambda^2 \geq 0; \quad s^x \geq 0; s^y \geq 0. \end{array} \right\}
$$

(7.16)

where x_j and y_j are the N-dimensional input and M-dimensional final output vectors, respectively, of DMU$_j$ ($j = 1, ..., J$). $\lambda = (\lambda 1, ..., \lambda J)$ is a J-dimensional vector of intensity variables. The potential slack in each input constraint, S_n^- ($n = 1, ..., N$) is normalized by its respective component of the directional vector $g^x = (g_1^x, ..., g_N^x)$, and the slacks in the output constraints, S_m^- ($m = 1, ..., M$), are normalized by the output directional vector, $g^y = (g_1^y, ..., g_N^y)$. The variables z are endogenous choice variables associated with the intermediate output. If $NSBI(.) = 0$, the DMU is efficient and inefficient otherwise.

7.8.1 R script for the NSBI model

The R script for the NSBI model is shown in Table 7.20.

7.8.2 Results for the NSBI model

The results for the NSBI model are based on the data set of Appendix E Table E.1.

The results from Table 7.21 present the inefficiency scores and the slacks for inefficient airlines. To explain how NSBI is derived from Eq. (7.16), using Qantas as an example,

$$
\text{NSBI}_{\text{Qantas}} = \frac{\frac{1}{N}\sum_{n=1}^{N}\frac{s_n^x}{g_n^x} + \frac{1}{M}\sum_{m=1}^{M}\frac{s_m^y}{g_m^y}}{2} = \frac{\frac{1}{2}\left(\frac{1182.946}{1} + \frac{3617506893}{1}\right) + \frac{1}{2}\left(\frac{1553589232}{1} + \frac{3986608998}{1}\right)}{2}
$$

$$
= 2289426576
$$

TABLE 7.20 R Script for Fukuyama and Weber (2010) NSBI model (excluding bad output).

```
## Load data (Reading from C drive and folder of data file)
data <- read.table(file="C:/Airline/NDEA1.txt", header=T,sep="\t")
attach(data)
X = cbind(x1,x2)
Y = cbind(y1,y2)
Z = cbind(z1,z2,z3)
gx = c(1,1)
gy = c(1,1)

dea.nsbi = function(xo,yo,Xref,Yref,Zref,gx,gy){

  library(lpSolve)
  n = nrow(Xref)
  p = ncol(Xref)
  q = ncol(Yref)
  r = ncol(Zref)

  obj.fun = 1/2*c(1/(p*gx),1/(q*gy),rep(0,n+n+r))
  constr.x = cbind(diag(p),matrix(0,p,q),t(Xref),matrix(0,p,n+r))
  constr.zx = cbind(matrix(0,r,p+q),-t(Zref),matrix(0,r,n),diag(r))
  constr.zy = cbind(matrix(0,r,p+q+n),t(Zref),-diag(r))
  constr.y = cbind(matrix(0,q,p),diag(q),matrix(0,q,n),-t(Yref),matrix(0,q,r))
  constr.lb = -diag(p+q+n+n+r)
  constr.lambda1 = cbind(matrix(0,1,p+q),matrix(1,1,n),matrix(0,1,n+r))
  constr.lambda2 = cbind(matrix(0,1,p+q+n),matrix(1,1,n),matrix(0,1,r))
  constr = rbind(constr.x,constr.zx,constr.zy,constr.y,constr.lb,constr.lambda1,constr.lambda2)
  constr.dir = c(rep("=",p),rep("<=",r+r),rep("=",q),rep("<=",p+q+n+n+r),"=","=")
  rhs = c(xo,rep(0,r+r),-yo,rep(0,p+q+n+n+r),1,1)

results = lp("max",obj.fun,constr,constr.dir,rhs)

  return(list(dea.nsbi=results$objval,slX=results$solution[1:p],slY=results$solution[(p+1):(p+q)]))
}
dea.nsbi.all = function(X,Y,Xref,Yref,Zref,gx,gy){
  m = nrow(X)
  n = nrow(Xref)
  p = ncol(Xref)
  q = ncol(Yref)
results = list(dea.nsbi=rep(0,m),slX=matrix(0,m,p),slY=matrix(0,m,q))
  for (i in 1:m){
    temp = dea.nsbi(X[i,],Y[i,],Xref,Yref,Zref,gx,gy)
    results$dea.nsbi[i] = round(temp$dea.nsbi,5)
    results$slX[i,] = round(temp$slX,5)
    results$slY[i,] = round(temp$slY,5)
  }
  return(results)
}
options(scipen = 999)
ineff_score<- dea.nsbi.all(X,Y,X,Y,Z,gx,gy)
data.frame(dmu, ineff_score)
```

TABLE 7.21 Results based on Fukuyama and Weber (2010) NSBI model (VRS).

```
R RStudio
File  Edit  Code  View  Plots  Session  Build  Debug  Profile  Tools  Help
  ▾  ▾        ▾        ▾    Go to file/function        ▾ Addins ▾

Console  Terminal ×
C:/Program Files/RStudio/NDEA/
> ineff_score <- dea.nsbi.arr(x,t,A,t,z,gx,gy)
> data.frame(dmu, ineff_score)
                           dmu          dea.nsbi      s1X.1          s1X.2       s1Y.1       s1Y.2
1                   Air Canada  554048491.03419      0.000        0.00000           0  2216193964
2    All Nippon Airways (ANA) 3002554135.92860      0.000  6043931792.39917  3616488959  2349795792
3            American Airlines        0.00013      0.000        0.00050           0           0
4              British Airways 2815694120.50671      0.000  7652843747.03489  1433296349  2176636386
5              Delta Air Lines 3219836734.85463   1731.000 11162000000.00000           0  1717345208
6                     Emirates        0.00001      0.000        0.00003           0           0
7             Garuda Indonesia        0.00000      0.000        0.00000           0           0
8                          KLM        0.00000      0.000        0.00000           0           0
9                    Lufthansa        0.00000      0.000        0.00000           0           0
10            Malaysia Airlines        0.00000      0.000        0.00000           0           0
11                      Qantas 2289426576.54546   1182.946  3617506893.65038  1553589232  3986608998
12   SAS Scandinavian Airlines        0.00000      0.000        0.00000           0           0
13           Singapore Airlines        0.00000      0.000        0.00000           0           0
14                          TAM 2046510890.01679      0.000        0.00000  3946798170  4239245390
15                 Thai Airways 8189408115.26851      0.000 21513008529.95710  6787076388  4457547543
16              United Airlines  232432780.07336      0.000        0.00000           0   929731120
```

Fukuyama and Weber (2010) showed that the NSBI can be decomposed into slack bias and network directional inefficiency (ND), formally expressed as

$$\text{NSBI}\,(x_o, y_o; g^x, g^y) = \text{SlackBias}(x_o, y_o; g^x, g^y) + \text{ND}(x_o, y_o; g^x, g^y) \quad (7.17)$$

The SlackBias in Eq. (7.17) gauges the underestimate of inefficiency that arises from ignoring input slacks and output slacks in the network framework.

Fukuyama and Weber (2010) also showed that NSBI can also be decomposed into NetBias and the black-box slacks-based inefficiency (SBI).

$$\text{NSBI}\,(x_o, y_o; g^x, g^y) = \text{NetBias}\,(x_o, y_o; g^x, g^y) + \text{SBI}\,(x_o, y_o; g^x, g^y) \quad (7.18)$$

Eq. (7.18) shows the bias in slacks-based inefficiency arising from neglecting a known network structure. For more details on the decomposition of NSBI of Eqs. (7.17) and (7.18), we direct readers to Fukuyama and Weber (2010).

The SlackBias and NetBias from Eqs. (7.17) and (7.18), respectively, can be derived by using the R script for *ND* and *SBI*. R script for *ND* is shown in Table 7.17 and the R script for *SBI* shown in Appendix E Table E.4. Results of SlackBias and NetBias under VRS and CRS conditions are presented in Tables 7.22 and 7.23, respectively.

The number of airlines exhibiting Slackbias >0 is greater under CRS as expected because efficiency scores based on CRS include scale efficiency. Airlines exhibiting Slackbias >0 suggest that these airlines have the opportunity to further expand at least one output or contract at least one input to improve efficiency.

TABLE 7.22 NSBI, ND and Slackbias (VRS and CRS).

	VRS			CRS		
	NSBI	ND	Slackbias	NSBI	ND	Slackbias
Air Canada	554,048,491	505	554,047,986	1,061,731,949	650	1,061,731,299
ANA All Nippon Airways	3,002,554,136	2358	3,002,551,778	3,378,190,350	2372	3,378,187,978
American Airlines	0	0	0	6,813,788,719	7576	6,813,781,143
British Airways	2,815,694,121	3042	2,815,691,079	3,513,331,509	3259	3,513,328,250
Delta Air Lines	3,219,836,735	2186	3,219,834,549	5,108,309,001	4022	5,108,304,979
Emirates	0	0	0	2,157,731,854	1778	2,157,730,076
Garuda Indonesia	0	0	0	388,235,075	25	388,235,050
KLM	0	0	0	169,214,876	2396	169,212,480
Lufthansa	0	0	0	9,563,727,692	19,099	9,563,708,593
Malaysia Airlines	0	0	0	518,391,555	2789	518,388,766
Qantas	2,289,426,577	4023	2,289,422,554	3,686,973,467	4178	3,686,969,289
SAS Scandinavian Airlines	0	0	0	672,304,192	1517	672,302,675
Singapore Airlines	0	0	0	0	0	0
TAM	2,046,510,890	2275	2,046,508,615	2,456,658,333	2358	2,456,655,975
Thai Airways	8,189,408,115	4367	8,189,403,749	8,886,352,871	4378	8,886,348,493
United Airlines	232,432,780	219	232,432,561	1,212,561,680	445	1,212,561,235

Source: Tables 7.19 and 7.21.

TABLE 7.23 NSBI, SBI and Netbias (VRS and CRS).

	VRS			CRS		
	NSBI	SBI	Netbias	NSBI	SBI	Netbias
Air Canada	554,048,491	49,438,223	504,610,268	1,061,731,949	470,608,510	591,123,439
ANA All Nippon Airways	3,002,554,136	2,347,433,803	655,120,333	3,378,190,350	2,709,083,002	669,107,348
American Airlines	0	0	0	6,813,788,719	4,226,408,876	2,587,379,843
British Airways	2,815,694,121	1,802,461,975	1,013,232,146	3,513,331,509	1,974,076,248	1,539,255,261
Delta Air Lines	3,219,836,735	2,976,499,683	243,337,052	5,108,309,001	3,399,918,128	1,708,390,873
Emirates	0	0	0	2,157,731,854	241,190,486	1,916,541,368
Garuda Indonesia	0	0	0	388,235,075	361,573,104	26,661,971
KLM	0	0	0	169,214,876	0	169,214,876
Lufthansa	0	0	0	9,563,727,692	7,968,461,713	1,595,265,979
Malaysia Airlines	0	0	0	518,391,555	418,646,623	99,744,932
Qantas	2,289,426,577	1,832,776,600	456,649,977	3,686,973,467	2,122,975,773	1,563,997,694
SAS Scandinavian Airlines	0	0	0	672,304,192	575,896,246	96,407,946
Singapore Airlines	0	0	0	0	0	0
TAM	2,046,510,890	1,406,435,711	640,075,179	2,456,658,333	1,737,783,537	718,874,796
Thai Airways	8,189,408,115	7,656,118,210	533,289,905	8,886,352,871	7,695,531,230	1,190,821,641
United Airlines	232,432,780	0	232,432,780	1,212,561,680	0	1,212,561,680

Source: Table 7.19 and SBI scores derived using Appendix E Table E.4 R script.

In Table 7.23, Netbias >0 suggests that airline operators can improve efficiency by examining the link between the nodes and not solely focus on changes in outputs and inputs.

7.9 A general network technology model to depict the airline provision–delivery model

The models before were based on a basic two-node NDEA model. That is, inputs x in 1 are used to produce intermediate products (z), and these intermediate products (z) are used in 2 to produce the final output y. Note that the previous models considered the following variables, 'flight personnel' and 'fuel burn' as intermediates, which is unrealistic in the context of our airline *provision–delivery* model. To construct an airline production model that is meaningful, we will now consider these two variables as inputs in node 2. Hence, the two-node NDEA model of Fig. 7.2 will now look as follows:

The model in Fig. 7.2 assumes a DMU employing inputs $x^1 = (x_1, ..., x_N)$ in node 1 producing intermediate product $z = (z_1, ..., z_Q)$, which is then used in node 2 alongside $x^2 = (x_1, ..., x_N)$, to produce final output $y = (y_1, ..., y_M)$. In other words, we formally define the production possibility set of the above network technology (NT) under VRS as follows:

$$NT = \left\{ (x^1, x^2, y) \, \middle| \, \begin{array}{l} x^1 \varepsilon \mathfrak{R}_+^N \text{ can produce } z \varepsilon \mathfrak{R}_+^Q \quad \text{(Node 1)} \\ (z, x^2) \varepsilon \mathfrak{R}_+^Q \text{ can produce } y \varepsilon \mathfrak{R}_+^M \quad \text{(Node 2)} \end{array} \right\}$$

$$= \left\{ (x^1, x^2, y) \, \middle| \, \begin{array}{c} x^1 \geq \sum_{j=1}^{J} x_j^1 \lambda_j^1; \; \sum_{j=1}^{J} z_j \lambda_j^1 \geq z; \\[2mm] \sum_{j=1}^{J} z_j \lambda_j^2 \leq z; \; x^2 \geq \sum_{j=1}^{J} x_j^2 \lambda_j^2; \; y \leq \sum_{j=1}^{J} y_j \lambda_j^2; \\[2mm] \sum_{j=1}^{J} \lambda_j^1 = 1; \; \sum_{j=1}^{J} \lambda_j^2 = 1; \\[2mm] z \geq 0; \; \lambda^1 \geq 0; \; \lambda^2 \geq 0. \end{array} \right\}$$

$$(7.19)$$

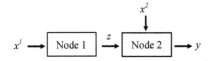

FIGURE 7.2 Airline *provision–delivery* two-node network DEA model.

7.9.1 R script for the NT model

The above NT model is a variation from Fukuyama and Weber (2010) NSBI model and hence based on a network slacks-based model. The R script for the NT model is shown in Table 7.24.

TABLE 7.24 R Script for NT model.

```
## Load data (Reading from C drive and folder of data file)
data <- read.table(file="C:/Airline/NDEA4.txt", header=T,sep="\t")
attach(data)

X1 = cbind(x1,x2)
X2 = cbind(x3,x4)
Y1 = cbind(y1,y2)
Z = as.matrix(z1)
gx1 = c(1,1)
gx2 = c(1,1)
gy1 = c(1,1)

gen.nsbi = function(xo1,xo2, yo2,X1,X2, Y1,Z,gx1,gx2, gy1){
  library(lpSolve)
  n = nrow(X1)
  p1 = ncol(X1)
  p2 = ncol(X2)
  q2 = ncol(Y1)
  r = ncol(Z)

  obj.fun = 1/3*c(1/(p1*gx1),1/(p2*gx2), 1/(q2*gy1),rep(0,n+n+r))

  constr.x1 = cbind(diag(p1),matrix(0,p1,p2+q2),t(X1),matrix(0,p1,n+r))
  constr.x2 = cbind(matrix(0,p2,p1),diag(p2),matrix(0,p2,q2+n),t(X2),matrix(0,p2,r))
  constr.zx = cbind(matrix(0,r,p1+p2+q2),-t(Z),matrix(0,r,n),diag(r))
  constr.zy = cbind(matrix(0,r,p1+p2+q2+n),t(Z),-diag(r))

  constr.y2 = cbind(matrix(0,q2,p1+p2),diag(q2),matrix(0,q2,n),-t(Y1),matrix(0,q2,r))
  constr.lambda1 = cbind(matrix(0,1,p1+p2+q2),matrix(1,1,n),matrix(0,1,n+r))
  constr.lambda2 = cbind(matrix(0,1,p1+p2+q2+n),matrix(1,1,n),matrix(0,1,r))
  constr.lb = -diag(p1+p2+q2+n+n+r)
  constr = rbind(constr.x1,constr.zx,
constr.x2,constr.zy,constr.y2,constr.lb,constr.lambda1,constr.lambda2)
  constr.dir = c(rep("<=",p1+p2),rep("<=",r),rep("<=",r),
rep("<=",q2),rep("<=",p1+p2+q2+n+n+r),"=","=")
  rhs = c(xo1,rep(0,r), xo2,rep(0,r),-yo2,rep(0,p1+p2+q2+n+n+r),1,1)

  results = lp("max",obj.fun,constr,constr.dir,rhs)
  return(results$objval)
}

gen.nsbi.all = function(X1,X2,Y1,X1ref,X2ref,Y1ref,Zref,gx1,gx2, gy1){
  m = nrow(X1)
  results = rep(0,m)
  for (i in 1:m){
    results[i] = round(gen.nsbi(X1[i,],X2[i,],Y1[i,],X1ref,X2ref,Y1ref,Zref,gx1,gx2,gy1))
  }
  return(results)
}
options(scipen = 999)
results<- gen.nsbi.all(X1,X2, Y1,X1,X2, Y1,Z,gx1,gx2,gy1)
data.frame(dmu, results)
```

> "=","=" refers to VRS. As noted in the previous line, constraint lambda1 and constraint lambda 2, these are equal to 1 in the next line (1,1).
>
> To change to CRS, both signs must be >=, and the values changed to 0 for both.

7.9.2 Results for the NT model

The results for the *NT* model are based on the data set of Appendix E Table E.5. The results from Tables 7.25 and 7.26 present the inefficiency scores where airline with an inefficiency score of 0 indicates the airline as efficient.

What we observe from the results shown in Tables 7.25 and 7.26 is that Singapore Airlines is efficient under both VRS and CRS, suggesting that it is also scale efficient. Recall that the ratio of CRS to VRS is the scale efficiency. In the context of network efficiency, Singapore Airlines is the only efficient airline suggesting that node 1 and node 2 of equation (7.19) must be efficient for the NT model to be efficient.

TABLE 7.25 Results from NT model (VRS).

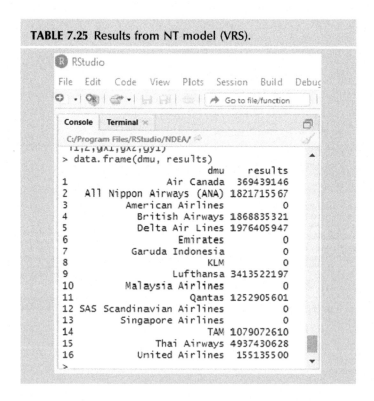

```
> data.frame(dmu, results)
                             dmu     results
1                     Air Canada   369439146
2         All Nippon Airways (ANA) 1821715567
3               American Airlines          0
4                 British Airways 1868835321
5                Delta Air Lines 1976405947
6                        Emirates          0
7               Garuda Indonesia          0
8                            KLM          0
9                      Lufthansa 3413522197
10              Malaysia Airlines          0
11                        Qantas 1252905601
12 SAS Scandinavian Airlines                0
13             Singapore Airlines          0
14                           TAM 1079072610
15                  Thai Airways 4937430628
16               United Airlines  155135500
>
```

TABLE 7.26 Results from NT model (CRS).

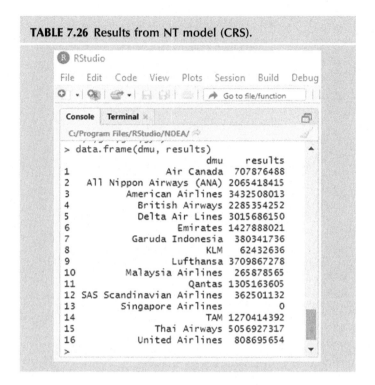

7.10 Conclusion

In this chapter, we covered the network models Kao and Hwang (2008) and Liang et al. (2008) centralized model, Farrell's efficiency model, the directional distance function model and the slack-based inefficiency model. The use of the *provision—delivery* model provided a better understanding on why the network model is more discriminatory than the standard 'black-box' DEA model in measuring inefficiency performance. The network DEA also provides insights into the inefficiencies within the *provision—delivery* model and identifying where inefficiencies lie. That said, this chapter only coveres a small number of NDEA models in the existing literature and its main purpose was to provide the reader a range of models on how to measure the network efficiency and inefficiencies. While the Kao and Hwang (2008) and Liang et al. (2008) centralized model provides a decomposition of the overall efficiency scores into node 1 and node 2 efficiency scores, the NTE, NRE and NCE models only provide overall efficiency scores. Likewise, the NDEA-NDDF and NSBI models only provide overall inefficiency scores. While not included in the R scripts, it is possible extend the R script for these models to include the decomposition of the overall scores into node 1 and node 2 scores. In addition, dynamic DEA can also be employed if one wishes to extend the NDEA over a period of time, although not covered in this book.

Appendix E

TABLE E.1 Data set for 'NDEA1.txt'.

DMU	Number of maintenance employees	Assets (property, plant and equipment) ('000 US$ PPP)	Available tonne kilometres (thousands) SATKTS	Flight personnel	Fuel burn (tonnes)	Passenger tonne kilometres performed (thousands) STKPTS	Freight and mail tonne kilometres performed (thousands) STKFTS
	x1	x2	z1	z2	z3	y1	y2
Air Canada	2293	7,217,121	13,028,613	8352	3,060,770	6,420,786	1,157,081
All Nippon Airways (ANA)	2591	14,651,828	14,683,532	6479	2,556,514	4,286,268	2,059,289
American Airlines	11,123	26,310,000	34,707,729	23,102	8,654,893	17,866,791	2,417,898
British Airways	4624	19,279,420	21,401,581	16,563	5,304,411	10,079,586	4,438,214
Delta Air Lines	6628	23,357,000	27,292,425	17,408	7,349,946	14,571,329	1,671,083
Emirates	3457	20,837,627	27,369,447	13,153	4,717,272	11,276,662	6,531,110
Garuda Indonesia	102	4,736,127	2,834,184	2187	676,347	1,514,745	282,129
KLM	4850	6,706,203	15,090,771	8101	3,027,818	7,347,192	4,093,466

Lufthansa	19,796	31,867,956	27,007,957	33,288	5,759,786	12,398,774	6,928,900
Malaysia Airlines	3762	3,953,020	7,292,543	5231	1,606,904	2,997,171	2,072,022
Qantas	6074	15,118,143	17,368,244	12,156	3,156,052	9,945,797	2,623,457
SAS Scandinavian Airlines	2047	2,954,620	4,152,670	4046	1,108,178	2,304,528	344,994
Singapore Airlines	438	22,323,127	21,286,125	9467	3,513,669	7,733,939	6,559,460
TAM	2789	8,314,066	7,840,248	6810	2,015,096	3,935,997	155,797
Thai Airways	4620	33,144,669	10,441,041	7374	2,417,856	4,725,671	2,157,255
United Airlines	4897	12,195,000	29,065,589	18,460	7,647,835	14,645,900	2,340,509

Source: Data from Appendix A Tables A.2 and A.3.

TABLE E.2 Data set for 'NDEA2.txt'.

DMU	Number of maintenance employees x1	Assets (property, plant and equipment) (US$ PPP) x2	Available tonne kilometres SATKTS z1	Flight personnel z2	Fuel burn (tonnes) z3	Passenger tonne kilometres performed STKPTS y1	Freight and mail tonne kilometres performed STKFTS y2	Average monthly wage of maintenance employees (US$ PPP) w1	Unit price of assets (US$ PPP) w2
Air Canada	2293	7,217,121,309	13,028,613,000	8352	3,060,770	6,420,786,000	1,157,081,000	1.2013	1.2013
All Nippon Airways (ANA)	2591	14,651,828,210	14,683,532,000	6479	2,556,514	4,286,268,000	2,059,289,000	115.1501	115.1501
American Airlines	11,123	26,310,000,000	34,707,729,000	23,102	8,654,893	17,866,791,000	2,417,898,000	1.0000	1.0000
British Airways	4624	19,279,419,885	21,401,581,000	16,563	5,304,411	10,079,586,000	4,438,214,000	0.7099	0.7099
Delta Air Lines	6628	23,357,000,000	27,292,425,000	17,408	7,349,946	14,571,329,000	1,671,083,000	1.0000	1.0000
Emirates	3457	20,837,626,952	27,369,447,000	13,153	4,717,272	11,276,662,000	6,531,110,000	2.0642	2.0642
Garuda Indonesia	102	4,736,126,766	2,834,184,000	2187	676,347	1,514,745,000	282,129,000	3007.0327	3007.0327
KLM	4850	6,706,202,766	15,090,771,000	8101	3,027,818	7,347,192,000	4,093,466,000	0.8482	0.8482

Lufthansa	19,796	31,867,955,706	27,007,957,000	5,759,786	12,398,774,000	6,928,900,000	0.8109	0.8109
Malaysia Airlines	3762	3,953,019,715	7,292,543,000	1,606,904	2,997,171,000	2,072,022,000	1.3329	1.3329
Qantas	6074	15,118,143,229	17,368,244,000	3,156,052	9,945,797,000	2,623,457,000	1.4415	1.4415
SAS Scandinavian Airlines	2047	2,954,620,474	4,152,670,000	1,108,178	2,304,528,000	344,994,000	8.9189	8.9189
Singapore Airlines	438	22,323,126,667	21,286,125,000	3,513,669	7,733,939,000	6,559,460,000	0.9005	0.9005
TAM	2789	8,314,065,642	7,840,248,000	2,015,096	3,935,997,000	155,797,000	1.2937	1.2937
Thai Airways	4620	33,144,669,141	10,441,041,000	2,417,856	4,725,671,000	2,157,255,000	11.8321	11.8321
United Airlines	4897	12,195,000,000	29,065,589,000	7,647,835	14,645,900,000	2,340,509,000	1.0000	1.0000

Source: Data from Appendix A Tables A.2 and A.3. Purchasing power parities (PPPs) drawn from World Bank's World Development Indicators (https://databank.worldbank.org/source/world-development-indicators#).

TABLE E.3 Data set for 'NDEA3.txt'.

DMU	Number of maintenance employees	Assets (property, plant and equipment) (US$ PPP)	Available tonne kilometres SATKTS	Flight personnel	Fuel burn (tonnes)	Passenger tonne kilometres performed STKPTS	Freight and mail tonne kilometres performed STKFTS	Unit price of airfare	Unit price of freight/mail
	$x1$	$x2$	$z1$	$z2$	$z3$	$y1$	$y2$	$p1$	$p2$
Air Canada	2293	7,217,121,309	13,028,613,000	8352	3,060,770	6,420,786,000	1,157,081,000	1.16,778,103	0.34057191
All Nippon Airways (ANA)	2591	14,651,828,210	14,683,532,000	6479	2,556,514	4,286,268,000	2,059,289,000	1.81816416	0.39663552
American Airlines	11,123	26,310,000,000	34,707,729,000	23,102	8,654,893	17,866,791,000	2,417,898,000	0.94810727	0.29851276
British Airways	4624	19,279,419,885	21,401,581,000	16,563	5,304,411	10,079,586,000	4,438,214,000	1.07212346	0.19942534
Delta Air Lines	6628	23,357,000,000	27,292,425,000	17,408	7,349,946	14,571,329,000	1,671,083,000	1.28072551	0.38608018
Emirates	3457	20,837,626,952	27,369,447,000	13,153	4,717,272	11,276,662,000	6,531,110,000	1.38766709	0.45537983
Garuda Indonesia	102	4,736,126,766	2,834,184,000	2187	676,347	1,514,745,000	282,129,000	3.38405284	1.22987523
KLM	4850	6,706,202,766	15,090,771,000	8101	3,027,818	7,347,192,000	4,093,466,000	0.85200933	0.31020841

Lufthansa	19,796	31,867,955,706	27,007,957,000	33,288	5,759,786	12,398,774,000	6,928,900,000	1.39843547	0.36721203
Malaysia Airlines	3762	3,953,019,715	7,292,543,000	5231	1,606,904	2,997,171,000	2,072,022,000	1.82720360	0.29419978
Qantas	6074	15,118,143,229	17,368,244,000	12,156	3,156,052	9,945,797,000	2,623,457,000	0.85934291	0.20920704
SAS Scandinavian Airlines	2047	2,954,620,474	4,152,670,000	4046	1,108,178	2,304,528,000	344,994,000	1.40462392	0.30131215
Singapore Airlines	438	22,323,126,667	21,286,125,000	9467	3,513,669	7,733,939,000	6,559,460,000	1.20947280	0.20141377
TAM	2789	8,314,065,642	7,840,248,000	6810	2,015,096	3,935,997,000	155,797,000	2.09929601	5.39972590
Thai Airways	4620	33,144,669,141	10,441,041,000	7374	2,417,856	4,725,671,000	2,157,255,000	2.73235833	0.91420916
United Airlines	4897	12,195,000,000	29,065,589,000	18,460	7,647,835	14,645,900,000	2,340,509,000	0.95569276	0.30446805

TABLE E.4 R script for Slacks-based inefficiency.

```
## Load data (Reading from C drive and folder of data file)
data <- read.table(file="C:/Airline/NDEA1.txt", header=T,sep="\t")
attach(data)
X = cbind(x1,x2)
Y = cbind(y1,y2)
gx = c(1,1)
gy = c(1,1)

dea.sbi = function(xo,yo,Xref,Yref,gx,gy,rts){

  library(lpSolve)
  n = nrow(Xref)
  p = ncol(Xref)
  q = ncol(Yref)
  obj.fun = 1/2*c(1/(p*gx),1/(q*gy),rep(0,n))  # objective function

  if (rts=="crs") {  # CRS case

    constr.x = cbind(diag(p),matrix(0,p,q),t(Xref))
    constr.y = cbind(matrix(0,q,p),diag(q),-t(Yref))
    constr.lb = -diag(p+q+n)
    constr = rbind(constr.x,constr.y,constr.lb)
    constr.dir = rep("<=",p+q+p+q+n)
    rhs = c(xo,-yo,rep(0,p+q+n))

  } else if (rts=="nirs") {

    constr.x = cbind(diag(p),matrix(0,p,q),t(Xref))
    constr.y = cbind(matrix(0,q,p),diag(q),-t(Yref))
    constr.lb = -diag(p+q+n)
    constr.lambda = cbind(matrix(0,1,p+q),matrix(1,1,n))
    constr = rbind(constr.x,constr.y,constr.lb,constr.lambda)
    constr.dir = rep("<=",p+q+p+q+n+1)
    rhs = c(xo,-yo,rep(0,p+q+n),1)

  } else {  # VRS case

    constr.x = cbind(diag(p),matrix(0,p,q),t(Xref))
    constr.y = cbind(matrix(0,q,p),diag(q),-t(Yref))
    constr.lb = -diag(p+q+n)
    constr.lambda = cbind(matrix(0,1,p+q),matrix(1,1,n))
    constr = rbind(constr.x,constr.y,constr.lb,constr.lambda)
    constr.dir = c(rep("<=",p+q+p+q+n),"=")
    rhs = c(xo,-yo,rep(0,p+q+n),1)

  }
  results = lp("max",obj.fun,constr,constr.dir,rhs)

  return(list(dea.sbi=results$objval,lambda=results$solution[(p+q+1):(p+q+n)],
       slX=results$solution[1:p],slY=results$solution[(p+1):(p+q)]))
}
dea.sbi.all = function(X,Y,Xref,Yref,gx,gy,rts){
  m = nrow(X)
  n = nrow(Xref)
  p = ncol(Xref)
  q = ncol(Yref)
  results = list(dea.sbi=rep(0,m),lambda=matrix(0,m,n),slX=matrix(0,m,p),slY=matrix(0,m,q))
```

TABLE E.4 R script for Slacks-based inefficiency.—cont'd

```
for (i in 1:m){
  temp = dea.sbi(X[i,],Y[i,],Xref,Yref,gx,gy,rts)
  results$dea.sbi[i] = round(temp$dea.sbi,5)
  results$lambda[i,] = round(temp$lambda,5)
  results$slX[i,] = round(temp$slX,5)
  results$slY[i,] = round(temp$slY,5)
  }
  return(results)
}
ineff_score<- dea.sbi.all(X,Y,X,Y,gx,gy,"vrs")
data.frame(dmu, ineff_score)
```

TABLE E.5 Data set for 'NDEA4.txt'.

DMU	Number of maintenance employees x1	Assets (property, plant and equipment) (US$ PPP) x2	Available tonne kilometres SATKTS z1	Flight personnel x3	Fuel burn (tonnes) x4	Passenger tonne kilometres performed STKPTS y1	Freight and mail tonne-k ilometres performed STKFTS y2
Air Canada	2293	7,217,121,309	13,028,613,000	8352	3,060,770	6,420,786,000	1,157,081,000
All Nippon Airways (ANA)	2591	14,651,828,210	14,683,532,000	6479	2,556,514	4,286,268,000	2,059,289,000
American Airlines	11,123	26,310,000,000	34,707,729,000	23,102	8,654,893	17,866,791,000	2,417,898,000
British Airways	4624	19,279,419,885	21,401,581,000	16,563	5,304,411	10,079,586,000	4,438,214,000
Delta Air Lines	6628	23,357,000,000	27,292,425,000	17,408	7,349,946	14,571,329,000	1,671,083,000
Emirates	3457	20,837,626,952	27,369,447,000	13,153	4,717,272	11,276,662,000	6,531,110,000
Garuda Indonesia	102	4,736,126,766	2,834,184,000	2187	676,347	1,514,745,000	282,129,000
KLM	4850	6,706,202,766	15,090,771,000	8101	3,027,818	7,347,192,000	4,093,466,000

Lufthansa	19,796	31,867,955,706	27,007,957,000	33,288	5,759,786	12,398,774,000	6,928,900,000
Malaysia Airlines	3762	3,953,019,715	7,292,543,000	5231	1,606,904	2,997,171,000	2,072,022,000
Qantas	6074	15,118,143,229	17,368,244,000	12,156	3,156,052	9,945,797,000	2,623,457,000
SAS Scandinavian Airlines	2047	2,954,620,474	4,152,670,000	4046	1,108,178	2,304,528,000	344,994,000
Singapore Airlines	438	22,323,126,667	21,286,125,000	9467	3,513,669	7,733,939,000	6,559,460,000
TAM	2789	8,314,065,642	7,840,248,000	6810	2,015,096	3,935,997,000	155,797,000
Thai Airways	4620	33,144,669,141	10,441,041,000	7374	2,417,856	4,725,671,000	2,157,255,000
United Airlines	4897	12,195,000,000	29,065,589,000	18,460	7,647,835	14,645,900,000	2,340,509,000

References

Castelli, L., Pesenti, R., Ukovich, W., 2004. DEA-like models for the efficiency evaluation of hierarchically structured units. European Journal of Operational Research 154 (2), 465–476.

Charnes, A., Cooper, W.W., Rhodes, E., 1978. Measuring the efficiency of decision-making units. European Journal of Operations Research 2, 429–444.

Chen, Y., Cook, W.D., Li, N., Zhu, J., 2009. Additive efficiency decomposition in two–stage DEA. European Journal of Operational Research 196, 1170–1176.

Chen, Y., Cook, W.D., Zhu, J., 2010. Deriving the DEA frontier for two-stage processes. European Journal of Operational Research 202, 138–142.

Fare, R., 1991. Measuring Farrell efficiency for a firm with intermediate inputs. Academia Economic Papers 19, 329–340.

Färe, R., Grosskopf, S., 2000. Network DEA. Socio-Economic Planning Sciences 34, 35–49.

Färe, R., Grosskopf, S., 1996. Productivity and intermediate products: a frontier approach. Economics Letters 50, 65–70.

Fukuyama, H., Matousek, R., 2011. Efficiency of Turkish banking: two-stage network system. Variable returns to scale model. Journal of International Financial Markets, Institutions and Money 21, 75–91.

Fukuyama, H., Matousek, R., 2017. Modelling bank performance: a network DEA approach. European Journal of Operational Research 259, 721–732.

Fukuyama, H., Weber, W.L., 2010. A slacks-based inefficiency measure for a two-stage system with bad outputs. Omega 38, 398–409.

Fukuyama, H., Weber, W.L., 2012. Estimating two-stage technology inefficiency. International Journal of Operations Research and Information Systems 3 (2), 1–23.

Halkos, G.E., Tzeremes, N.G., Kourtzidis, S.A., 2014. A unified classification of two-stage DEA models. Surveys in Operations Research and Management Science 19 (1), 1–16.

Kao, C., Hwang, S.N., 2011. Decomposition of technical and scale efficiencies in two-stage production systems. European Journal of Operational Research 211, 515–519.

Kao, C., Hwang, S.N., 2014. Multi-period efficiency and Malmquist productivity index in two-stage production systems. European Journal of Operational Research 232, 512–521.

Kao, C., Hwang, S.N., 2008. Efficiency decomposition in two-stage data envelopment analysis: an application to non-life insurance companies in Taiwan. European Journal of Operational Research 185 (1), 418–429.

Liang, L., Cook, W.D., Zhu, J., 2008. DEA models for two-stage processes: game approach and efficiency decomposition. Naval Research Logistics 55, 643–653.

Lim, D-J, 2021. Distance Measure Based Judgment and Learning. Package 'DJL'. Version 3.7. https://cran.rstudio.com/web/packages/DJL/index.html.

Seiford, L.M., Zhu, J., 1999. Profitability and marketability of the top 55 US commercial banks. Management Science 45 (9), 1270–1288. https://doi.org/10.1287/mnsc.45.9.1270.

Tone, K., Tsutsui, M., 2009. Network DEA: a slacks-based measure approach. European Journal of Operational Research 197 (1), 243–252.

Chapter 8

Sources of airline performance

8.1 Introduction

Chapters 3–7 examined a range of DEA-type models on measuring the efficiency and productivity of airlines. However, DEA-type analysis has long been criticized as they have no statistical significance and are deterministic and unable to determine sources of efficiency. Further more, DEA assumes that DMUs have full control over their inputs, suggesting that such variables are all discretionary. Ouellette and Vierstraete (2004) argued that nondiscretionary inputs are present in virtually all sectors and that they need to be incorporated into any DEA model. The handling of nondiscretionary variables is discussed in Banker and Morey (1986), Ray (1991), Ruggiero (1996, 1998), Mũniz (2002), Nemoto and Goto (2003), Bilodeau et al. (2004), Ouellette and Vierstraete (2004) and Essid et al. (2010). Of these, two basic approaches are identified with regards to handling nondiscretionary inputs. The first approach directly incorporates nondiscretionary inputs in the DEA program. Studies such as Banker and Morey (1986) and Ruggiero (1996) employ this approach. The second approach omits the nondiscretionary inputs from the initial DEA analysis and only incorporates them in sequential non-DEA stages. Studies include Ray (1991), Mũniz (2002) and Simar and Wilson (2007).

In the context of the airline industry, inputs can be distinguished into two types—endogenous and exogenous. Endogenous inputs are physical inputs that are used directly to the production of outputs. From our previous chapters, we observe that the inputs in the *delivery* model, flight personnel, ATK and fuel burn are discretionary in nature, meaning that management has control over them. Exogenous inputs are external factors that can influence the efficiency of a DMU, for example, ownership, education level, location, poor weather or a strike action. These examples of exogenous factors are nondiscretionary in nature, and while DMUs have no control over nondiscretionary factors in DEA, they should still be factored into the efficiency analysis to ascertain their impact on the efficiencies of DMUs (Golany and Roll, 1993). To determine the impact exogenous factors have on the efficiency of a DMU, a second-stage analysis is thus performed by regressing the DEA efficiency scores against the exogenous factors. While there are other approaches in which DEA can incorporate nondiscretionary variables in its linear

Productivity and Efficiency Measurement of Airlines. https://doi.org/10.1016/B978-0-12-812696-7.00007-6

programming (LP), Ruggiero's (1998) simulation showed that the two-stage DEA outperforms the one-stage DEA models and also supports Ray (1988). They showed that including nondiscretionary inputs in the LP model for DEA amounts to an assumption of free disposability of these inputs, which is unrealistic.

To overcome the issue of the lack of statistical significance and inability to determine sources of efficiency, and given the discretionary and nondiscretionary inputs airlines face, this chapter employs the two-stage analysis as suggested by Ray (1991) and Coelli et al. (2005). Second-stage analysis has been discussed extensively in Coelli et al. (1999), Cooper et al. (2000), and Hoff (2007) and employed in numerous DEA studies with the vast majority using ordinary least squares (OLS) regression, generalized least squares (GLS) regression, Tobit regression and Simar and Wilson's (2007) bootstrap truncated regression model. We will describe the OLS, GLS, Tobit and Simar and Wilson's (2007) models in this chapter.

This chapter is divided into five sections. Following the introduction, in Section 8.2, we describe the data to be used in the second-stage regression. Section 8.3 describes two tests that need to be performed to ensure reliability of results. These are the multicollinearity test and separability test. Section 8.4 presents the OLS estimator. In Section 8.5, we describe the GLS estimator. Section 8.6 describes the Tobit regression model (TRM). Section 8.7 describes Simar and Wilson (2007) bootstrap truncated regression model. This chapter concludes with some brief comments in Section 8.8.

8.2 Data for second-stage regression

As previously described in Chapter 3, the airline production model comprises *provision* and *delivery* models. To demonstrate the statistical analysis and determine sources of efficiency, we will focus on the *delivery* model. The data for the *delivery* model for second stage regression analysis comprise environmental variables, which are both discretionary and nondiscretionary (comprising operational and organizational factors) in nature. Similar to Barbot et al. (2008), which considered internal (i.e., operational factors) and external conditions (i.e., organizational factors), our study also incorporates these nondiscretionary factors, which thus supports our choice of determinants. However, even when technically discretionary, they may not be amenable to change in the short run. We therefore expect all of these variables to have some impact on airline efficiency even though they are not included in the input—output specification. We include the following environmental variables. First, the average stage length or average stage distance flown per aircraft is measured by dividing the aircraft kilometres flown by the related number of aircraft departures. This variable was used in Merket and Hensher (2011). Second, weight load factor (WLF) is an indicator of the ability of firms to behave efficiently in light of external market pressure (Bhadra, 2009). Third, operational ratio in terms of revenue divided by expenses 'may influence airlines' technical efficiencies in ways that are not captured in determining the magnitude of the technical efficiencies. Operating under a high cost environment, generally speaking, may induce higher technical efficiencies out of the resource use' (Bhadra, 2009, p. 233). The data to perform the second-stage regression are presented in Appendix F Table F.1. The efficiency scores are based on the CCR DEA model of Chapter 3 as shown in Table 3.10.

8.3 Multicollinearity test and separability test

To derive reliable and meaningful results, the test for multicollinearity needs to be performed. As the study employs a second-stage regression analysis, the assumption of separability needs to be tested (Daraio et al., 2018).

8.3.1 R script for multicollinearity test

The R script for the multicollinearity test is shown in Table 8.1 drawn from the R package 'mctest' developed by Imdadullah and Aslam (2017). The 'mctest' R package provides some widely used collinearity diagnostic measures namely, the R-squared value, variance inflation factor (VIF), tolerance limit (TOL), eigenvalues, condition number (CN) and condition index (CI), Leamer's method, Kleins rule, three tests proposed by Fernandes et al. (2018), the Red indicator and Theil's measure. As noted in Imdadullah et al. (2016, p. 496), 'since no specific collinearity diagnostic measure is superior and each of these measures has different collinearity detection criterion (threshold value) proposed by various authors in the textbooks and research articles, there is need to study multiple collinearity diagnostics'. Hence, the reason for regression analysis to rely on more than one collinearity diagnostic measures. For more details of these collinearity diagnostic measures in 'mctest' and the problems faced in multicollinearity, we direct readers to Imdadullah et al. (2016).

TABLE 8.1 R script for multicollinearity test.

```
## load mctest
library(mctest)

## Load data
data <- read.table(file="C:/Airline/Delivery_regression1.txt", header=T,sep="\t")
attach(data)
model <- lm(y ~ x1+x2+x3+x4, data=as.data.frame(data))

## Overall diagnostic measures with intercept term
omcdiag(model, Inter=TRUE)

## Overall diagnostic measures without intercept term
omcdiag(model, Inter=FALSE)

## Individual diagnostic measures
imcdiag(model, corr=TRUE)

## Individual diagnostic measures (0 - collinearity is not detected) or 1 - collinearity is
detected)
imcdiag(model, all=TRUE)
```

8.3.2 Interpretation of the multicollinearity test results

Tables 8.2A, 8.2B and 8.2C present the results from the multicollinearity test. Results from the omcdiag (Table 8.2A) show that all of the overall collinearity diagnostic measures do not detect the presence of multicollinearity among regressors, apart from the CN when we include the intercept term. In terms of the results from the imcdiag (Tables 8.2B and 8.2C), the individual multi-collinearity diagnostics reveal that all of the measures do not detect the presence of multicollinearity among regressors.

TABLE 8.2A Multicollinearity test results.

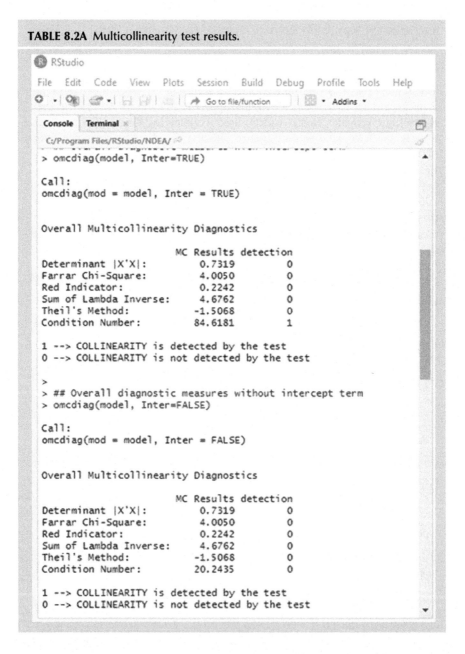

TABLE 8.2B Multicollinearity test results.

```
R  RStudio

File   Edit   Code   View   Plots   Session   Build   Debug   Profile   Tools   Help

  ◎ ▾ | 🖸 | 🖅 ▾ | 🗎 🗎 | 🖺 | → Go to file/function    | 🔡 ▾ Addins ▾

 Console   Terminal ×

 C:/Program Files/RStudio/NDEA/
 >
 > ## Individual diagnostic measures
 > imcdiag(model, corr=TRUE)

 Call:
 imcdiag(mod = model, corr = TRUE)


 All Individual Multicollinearity Diagnostics Result

         VIF    TOL     Wi     Fi Leamer   CVIF Klein   IND1   IND2
 x1 1.2694 0.7878 1.0776 1.7510 0.8876 3.8416     0 0.1969 1.5016
 x2 1.2040 0.8306 0.8159 1.3259 0.9114 3.6436     0 0.2076 1.1987
 x3 1.1189 0.8937 0.4757 0.7730 0.9454 3.3862     0 0.2234 0.7521
 x4 1.0839 0.9226 0.3355 0.5453 0.9605 3.2802     0 0.2307 0.5476

 1 --> COLLINEARITY is detected by the test
 0 --> COLLINEARITY is not detected by the test

 x1 , x2 , x4 , coefficient(s) are non-significant may be due to multicollinearity

 R-square of y on all x: 0.6907

 * use method argument to check which regressors may be the reason of collinearity
 ===================================

 Correlation Matrix
           x1          x2          x3          x4
 x1  1.0000000 -0.36431513 -0.2474925 -0.18014757
 x2 -0.3643151  1.00000000  0.2344937 -0.07634894
 x3 -0.2474925  0.23449366  1.0000000 -0.11992276
 x4 -0.1801476 -0.07634894 -0.1199228  1.00000000
```

TABLE 8.2C Multicollinearity test results.

```
R  RStudio

File   Edit   Code   View   Plots   Session   Build   Debug   Profile   Tools   Help

  ◎ ▾ | 🖸 | 🖅 ▾ | 🗎 🗎 | 🖺 | → Go to file/function    | 🔡 ▾ Addins ▾

 Console   Terminal ×

 C:/Program Files/RStudio/NDEA/
 > ## Individual diagnostic measures (0 - collinearity is not detected) or 1 - collinearity
   is detected)
 > imcdiag(model, all=TRUE)

 Call:
 imcdiag(mod = model, all = TRUE)


 All Individual Multicollinearity Diagnostics in 0 or 1

    VIF TOL Wi Fi Leamer CVIF Klein IND1 IND2
 x1   0   0  0  0      0    0     0    0    0
 x2   0   0  0  0      0    0     0    0    0
 x3   0   0  0  0      0    0     0    0    0
 x4   0   0  0  0      0    0     0    0    0

 1 --> COLLINEARITY is detected by the test
 0 --> COLLINEARITY is not detected by the test

 x1 , x2 , x4 , coefficient(s) are non-significant may be due to multicollinearity

 R-square of y on all x: 0.6907

 * use method argument to check which regressors may be the reason of collinearity
 ===================================
 >
 > |
```

8.3.3 R script for separability test

The R script for the separability test is shown in Table 8.3 using Wilson (2008) FEAR software package version 3.1 based on R package. Following Daraio et al. (2018), bootstrap replications of 1000 and 10 splits are performed for the separability test. The tests present P-values based on the bootstrap method described in Simar and Wilson (2020). If the results reveal that P-values are greater than 0.05, we do not reject the null for the alternative, which suggests that the separability assumption holds.

As our data set is continuous, we use the command 'test.sep.cont'. If the data set contains discrete values such as dummy variables, then the command 'test.sep.disc' must be used. However, if the data set comprises both continuous and discrete, then one has to perform the continuous and discrete test separately.

TABLE 8.3 R script for separability test.

```
## SEPARABILITY TEST (CONTINUOUS DATA).
library(FEAR)

## load data
data <- read.table(file="C:/Airline/Delivery_regression2.txt", header=T,sep="\t")

attach(data)

X=t(matrix(c(data$x1,data$x2,data$x3),nrow=16,ncol=3))
Y=t(matrix(c(data$y1,data$y2),nrow=16,ncol=2))
Z=t(matrix(c(data$z1,data$z2,data$z3),nrow=16,ncol=3))

## Estimator - 1 for the FDH estimator, 2 (default) for the VRS-DEA estimator, 3 for the CRS-DEA
estimator
## Orientation - 1 for input orientation, 2 for output orientation, or 3 for hyperbolic graph
orientation.

test.sep.cont(X, Y, Z, ESTIMATOR = 2, ORIENTATION = 1, METRIC = 1, NSPLIT = 10,
NREP = 1000, errchk = TRUE)
```

8.3.4 Interpretation of the separability test results

Table 8.4 presents the results from the separability test. The results reveal that P-values (0.142 and 0.983) are greater than 0.05, and we do not reject the null for the alternative, which suggests that the separability assumption holds.

TABLE 8.4 Separability test results.

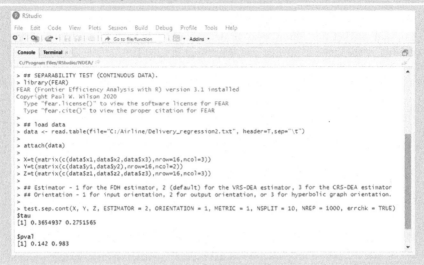

8.4 Ordinary least squares regression model

We start off the regression analysis by describing the ordinary least squares (OLS). OLS has been widely used in second-stage regression with efficiency scores as the dependent variable and the explanatory variables as independent variables. While the extant literature shows that OLS has been used in a range of studies, it has not been employed in airline efficiency analysis.

8.4.1 R script for ordinary least squares

The R script for the second-stage regression is shown in Table 8.5 and is based on the 'Benchmarking' package developed by Bogetoft and Otto (2011, 2015).

TABLE 8.5 R script for OLS.

```
library(Benchmarking)
library(AER)
## Load data
d <- read.table(file="C:/Airline/Delivery_regression2.txt", header=T,sep="\t")

x <- cbind(d$x1, d$x2, d$x3)
y <- cbind(d$y1, d$y2)
e <- dea(x,y,RTS="crs",ORIENTATION="in")

E <- eff(e)
eOLS <- lm(E ~ z1+z2+z3, data=d)

summary(eOLS)
```

8.4.2 Interpretation of results

The results from the OLS regression shown in Table 8.6 show that WLF (z2) is statistically significant and positively impacts on efficiency suggesting that external market conditions does bring about efficiency within airlines. This finding is the same as in Lee and Worthington (2014).

TABLE 8.6 OLS regression results.

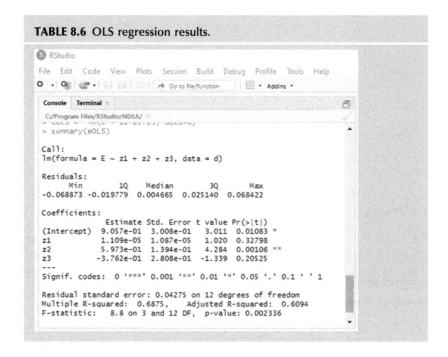

8.5 Generalized least squares regression model

The debate on the validity of using OLS in the second-stage regression is widespread in the extant literature. McDonald (2009, p. 797) argued that under ideal conditions, 'OLS is a consistent estimator, and, if White's heteroskedastic-consistent standard errors are calculated, tests can be performed which are valid for a range of disturbance distribution assumptions'. However, OLS works well when the assumptions are not violated, that is, the OLS regression assumes that the error variances are homoskedastic errors, uncorrelated and also normally distributed. However, violation of the 'Best Linear Unbiased Estimates' assumption of Gauss Markov theorem results in having serial correlation and homoskedasticity assumption violation, which leads to unreliable confidence intervals and hypotheses tests, yielding biased and inconsistent parameter estimates (Pindyck and Rubinfeld, 1991). To overcome this problem, one could use the generalized least squares (GLS) estimator as it is capable of producing estimates that are 'Best Linear Unbiased Estimates'. The GLS estimator is thus unbiased, consistent, efficient and asymptotically normal. GLS has been widely used in the second-stage regression analysis of airline efficiency. Such studies include Pitfield et al. (2012), Saranga and Nagpal (2016) and Mhlanga et al. (2018). The GLS estimator is expressed as follows:

$$Y_i = \alpha + \beta X_i + \varepsilon_i \qquad (8.1)$$

where Y_i is the efficiency score of airline i, α is the intercept, β is the coefficient for the environmental variable X and ε_i is the error term.

8.5.1 R script for generalized least squares

The GLS R script for the second-stage regression is shown in Table 8.7. The GLS R script was developed by Pinheiro et al. (2022) under the 'nlme' package version 3.1-160 available at https://cran.r-project.org/web/packages/nlme/nlme.pdf.

TABLE 8.7 R script for GLS.

```
library(nlme)
## Load data
data <- read.table(file="C:/Airline/Delivery_regression1.txt", header=T,sep="\t")

flm1 <- gls(y ~ z1+z2+z3, data)
flm2<-update(flm1, weights = varPower())
summary(flm1)
summary(flm2)
```

8.5.2 Interpretation of results

Two sets of results were generated from the R script, flm1 and flm2. Results of flm1 shown in Table 8.8 are based on the assumption when the error variances are homoskedastic. Hence, the results from the GLS regression would be identical to the OLS results in Table 8.6, which shows that the GLS is a generalization of the OLS where the errors are uncorrelated and have equal variances.

The GLS estimator yields efficient estimates by weighting individuals in terms of the size of their variances and whether their errors are correlated or not. To include the weights into our GLS estimator, we include an additional line 'result2<-update(result1, weights = varPower())' in the R script to instruct R of the weights in terms of the size of their variances. Table 8.9 shows the results of flm2 when we assume the presence of heteroskedasticity.

Table 8.9 shows that WLF (z2) is statistically significant and positively impacts on efficiency suggesting that external market conditions do bring about efficiency within airlines. This finding is the same as in Lee and Worthington (2014).

TABLE 8.8 GLS regression for flm1.

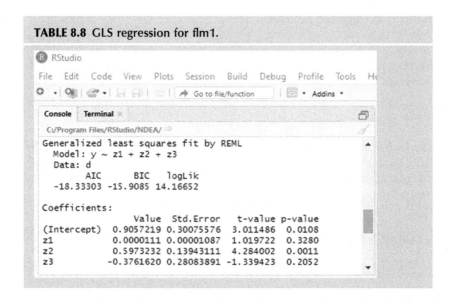

TABLE 8.9 GLS regression for flm2.

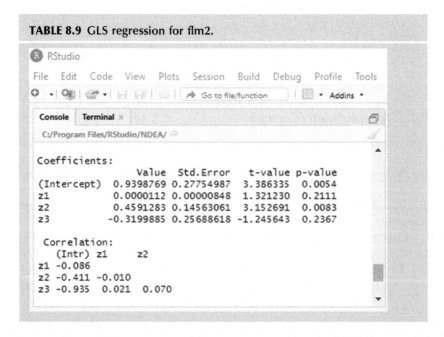

8.6 Tobit regression model

The Tobit regression model (also known as a censored regression model) has been widely used to determine sources of airline efficiency. Studies that have used the Tobit model include Scheraga (2004), Chiou and Chen (2006), Bhadra (2009), Merket and Hensher (2011), Saranga and Nagpal (2016) and Mhlanga (2019). The Tobit regression model (also known as a censored regression model) estimates linear relationships between variables and is employed when the dependent variable is bounded from below or above or both. The Tobit regression is expressed as

$$Y_i^* = Z_t\beta + u_t$$
$$Y_i = 0 \text{ if } Y \leq 0; \quad Y_i = Y_i^* \text{ if } 0 < Y_i^* \leq 1; \quad Y_i = 1 \text{ if } Y_i^* > 1 \tag{8.2}$$

where Z_t is the vector of independent variables, β is a vector of unknown coefficients and u_t is the independently distributed error term assumed to be normal with a zero mean and constant variance $N(0, \sigma^2)$. In the context of DEA, the efficiency estimate from DEA is the dependent variable and is regressed against nondiscretionary environment factors. As the DEA technical efficiency estimate lies between 0 and 1, it is thus censored from the left (i.e., from below) as well as from the right (i.e., from above). The technical efficient function of airline i efficiency can thus be written as

$$E_i = \alpha + \beta_1 + u_t$$

where E_i is the DEA technical efficiency score for airline i, α is a constant term, β_1 is the coefficient of the environmental variable (i.e., independent variables) and u_t is an error term.

The Tobit regression model is preferred over the OLS and GLS regression because of the censored dependent variable of efficiency scores of firms (Hsiao et al., 2010; Kao et al., 2011; Barth et al., 2013; Jha et al., 2013; Lee and Chih, 2013a,b). This is because the technical efficiency scores lie between 0 and 1. For detailed discussion on the Tobit regression, we direct readers to Hoff (2007) and Bogetoft and Otto (2011). McDonald (2009), however, argues in favor of OLS over Tobit because the latter's efficiency scores are fractional data and are an inconsistent estimator, which is not generated by the censoring process.

8.6.1 R script for the Tobit regression

The R script for the Tobit regression model is presented in Table 8.10. The function tobit used to perform the tobit regression is part of the AER (Applied Econometrics with R) package version 1.2−10 developed by Kleiber and Zeileis (2020).

TABLE 8.10 R script for Tobit regression.

```
library(Benchmarking)
library(AER)
## Load data
d <- read.table(file="C:/Airline/Delivery_regression2.txt", header=T,sep="\t")

x <- cbind(d$x1, d$x2, d$x3)
y <- cbind(d$y1, d$y2)
e <- dea(x,y,RTS="crs",ORIENTATION="in")

E <- eff(e)

eTob <- tobit(E ~ z1+z2+z3, left=0, right=1, data=d)
summary(eTob)
```

8.6.2 Interpretation of Tobit results for the 'delivery model'

The Tobit results are shown in Table 8.11. The results are similar to the GLS findings that WLF (z2) is statistically significant and positively impacts on efficiency suggesting that external market conditions does bring about efficiency within airlines.

8.7 Simar and Wilson (2007) regression model

Simar and Wilson (2007) criticized the conventional two-stage DEA approaches of OLS [and GLS though not mentioned in Simar and Wilson (2007)] and Tobit regression because the DEA efficiency estimates are, by construction, serially correlated, therefore invalidating conventional inferences in the second stage. Furthermore, the studies that employ the conventional regression models do not describe a coherent data generating process. To resolve this problem, Simar and Wilson (2007) proposed the bootstrap truncated regression comprised of two complementary consistent procedures in the two-stage DEA approach: algorithm 1 (the single bootstrap truncated regression) and algorithm 2 (the double-bootstrap DEA). Algorithm 2 is preferred over algorithm 1 (see Fernandes et al. (2018) and Simar and Wilson (2007) for detailed discussions). The double-bootstrap procedure enables a consistent inference within models explaining efficiency scores, while simultaneously producing standard errors and confidence intervals for these efficiency scores (Simar and Wilson, 2007, 2011, 2013). Simar and Wilson's (2007) bootstrap truncated regression model has been widely used to determine sources of airline efficiency. Studies include Barros and Peypoch (2009), Assaf and Josiassen (2011), Wu et al. (2013), Lee and Worthington (2014) and Choi (2017).

TABLE 8.11 Tobit regression results.

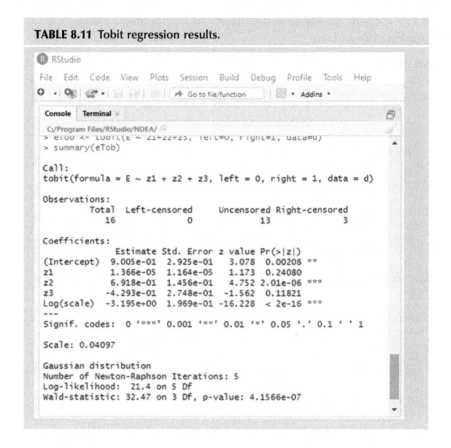

By combining DEA with bootstrapping technique, a set of bias-corrected DEA efficiency scores (denoted $\widehat{\widehat{\theta}}_i$) and confidence intervals are derived and then regressed on a set of hypothesized environmental factors using the following regression model:

$$\widehat{\widehat{\theta}}_i = a + Z_i \delta + \varepsilon_i, \quad i = 1, ..., n \qquad (8.3)$$

where $\varepsilon_i \sim N(0, \sigma_\varepsilon^2)$ with left truncation at $1 - Z_i \delta$; a is a constant term and Z_i is a vector of specific environmental variables for firm i that is expected to affect the efficiency of firm performance. The double-bootstrapping truncated regression step-by-step description is outlined in several studies (Simar and Wilson, 2007; Barros and Assaf, 2009; Barros and Garcia-del-Barrio, 2011). For brevity sake, we omit the description here and direct readers to these papers.

8.7.1 R script for Simar and Wilson (2007) double-bootstrap truncated regression

We use the R script 'Package rDEA' version 1.2-6 developed by Simm and Besstremyannaya (2016) to carry out the DEA and double-bootstrap estimations and present it in Table 8.12.

TABLE 8.12 R script for Simar and Wilson (2007) double-bootstrap truncated regression.

```
## Load library(rDEA)
library(rDEA)

## load data
data <- read.table(file="C:/Airline/Delivery_regression2.txt", header=T,sep="\t")
attach(data)

##Y refers to outputs, X refers to inputs, Z refers to environmental variables
##Note: If costmin is used, then need a fourth matrix called W <- cbind(w1, ...)

Y <- cbind(y1,y2)
X <- cbind(x1,x2,x3)
Z <- cbind(z1,z2,z3)

## Naive input-oriented DEA score for all firms under constant returns-to-scale
firms=1:16

## Bias-corrected DEA score in input-oriented model with environmental variables,
constant returns-to-scale
di_env=dea.env.robust(X=X[firms,], Y=Y[firms,], Z=Z[firms,], model="input",
RTS="constant", L2=2000, alpha=0.05)

## The following is the vector of robust reciprocal of DEA score (after the second loop).
di_env$delta_hat_hat

## The matrix of lower and upper bounds for beta using alpha confidence intervals (beta
is robust coefficients in the truncated regression of the reciprocal of DEA score on
environmental variables)
di_env$beta_ci

## The following is the vector of robust coefficients in the truncated regression of
reciprocal of DEA score on environmental variables (after the second loop)

## NOTE: A positive sign in coefficient indicates a negative influence on efficiency,
while a negative sign indicates a positive influence. This is because the dependent
variable is the reciprocal of the efficiency score, and is thus larger than or equal to one

di_env$beta_hat_hat

## The matrix of lower and upper bounds for sigma using alpha confidence intervals
(sigma is the robust standard deviation in the truncated regression of the reciprocal of
DEA score on environmental variables)
di_env$sigma_ci
```

8.7.2 Interpretation of Simar and Wilson's (2007) double-bootstrap truncated regression results

The main results from the R script 'Package rDEA' are in the sections that show the 'di_env$beta_ci' and the 'di_env$beta_hat_hat'. The former refers to the 95% confidence intervals for each of the environmental variable as shown in Table 8.13. The latter refers to the derived coefficient for each of the environmental variable.

From the observed coefficients of each of the environmental variable, we note that z2 (WLF) is statistically significant and positively impacts on efficiency suggesting that external market conditions do bring about efficiency within airlines. It is important to note that for the coefficients reported in Table 8.13, a positive sign in coefficient is shown, which indicates a negative influence on efficiency, while a negative sign indicates a positive influence. This is noted in Simm and Besstremyannaya (2016) that this is due to the dependent variable being the reciprocal of the efficiency score and is thus larger than or equal to one. Our results also provide an interesting outcome showing z1 (average stage length) as significant and negatively impacts efficiency, which has the same finding as Merket and Hensher (2011) and Huang

TABLE 8.13 Simar and Wilson's (2007) double-bootstrap truncated regression results.

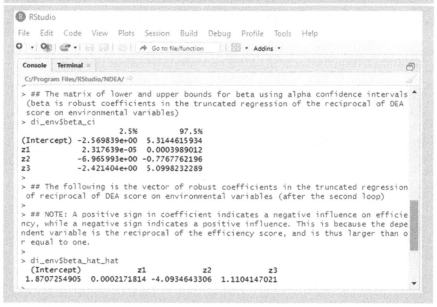

et al. (2021). Under OLS and Tobit, these two models showed average stage length as insignificant, which suggests the robustness of the double-bootstrap approach.

8.8 Conclusion

In this chapter, we first covered two important tests that should perform first to derive meaningful results. First, the multicollinearity test should be performed to determine if the independent variables in the regression model are highly correlated to each other. High correlation creates an overfitting problem and makes it difficult to interpret the model. Second, the separability test should be performed to determine whether environmental variables (normally nondis-cretionary in nature) can be separated and regressed as independent variables or whether they should be included as inputs in the first-stage efficiency analysis. The chapter covered three widely used regression models, GLS, Tobit regression and Simar and Wilson's (2007) bootstrap truncated regression, to determine sources of efficiency. As noted in Simar and Wilson (2007), studies that employ the conventional regression models do not describe a coherent data generating process. In addition, the DEA efficiency estimates via the conventional two-stage DEA approaches are, by construction, serially corre-lated and are therefore invalidating conventional inferences in the second stage. The results from the airline performance efficiency analysis from all three regression methods show similarities in identifying WLF as a significant variable that positively impacts airline efficiency. Simar and Wilson (2007) was the only regression model that identified 'average stage length' as sig-nificant and negatively impacts efficiency, demonstrating the robustness of Simar and Wilson's double-bootstrap truncated regression approach. Notwithstanding, Banker et al. (2019) performed simulations using OLS, Tobit and Simar and Wilson's (2007) bootstrap truncated regression and found that the OLS and Tobit performed as well as, if not better than Simar and Wilson's (2007) bootstrap truncated regression. As such, because each of these models are based on set assumptions, Banker et al. (2019) suggest that the choice of model should be determined by its appropriateness and research context.

Appendix F

See Table F.2.

TABLE F.1 Dataset for 'Delivery_regression1', 2009.

DMU	Efficiency y	Average stage length z1	Weight load factor z2	Operational ratio z3
Air Canada	0.9097854	2351.03754	0.581632673	0.968572849
All Nippon airways (ANA)	0.7294383	1113.020502	0.432154675	1.005479454
American airlines	0.9281796	2033.769025	0.584442993	0.944786721
British airways	0.9096533	2368.813145	0.678351754	0.95344436
Delta air lines	0.9887013	2078.429466	0.595125277	0.967574377
Emirates	0.9779102	4175.382525	0.650644202	1.024964862
Garuda Indonesia	0.9333141	1156.502238	0.634000474	1.050834426
KLM	1	1877.147973	0.758122829	0.958613422
Lufthansa	0.9447873	1321.710521	0.715628879	1.010927382
Malaysia airlines	0.9765008	2183.9869	0.69512007	0.956114589
Qantas	1	1636.73531	0.723691698	1.015279346
SAS Scandinavian airlines	0.9691051	815.6428041	0.638028545	0.934966136
Singapore airlines	1	4646.417801	0.67148901	0.960890949
TAM	0.8766804	1201.092723	0.521895991	1.034039484
Thai airways	0.8818556	2579.775988	0.659218367	1.06012395
United airlines	0.9356125	2327.170577	0.584416473	0.978280543

TABLE F.2 Dataset for 'Delivery_regression2', 2009.

DMU	Flight personnel x1	Available tonne-kilometres (thousands) x2	Fuel burn (tonnes) x3	Passenger tonne-kilometres performed (thousands) y1	Freight and mail tonne-kilometres performed (thousands) y2	Average stage length z1	Weight load factor z2	Operational ratio z3
Air Canada	8352	13,028,613	3,060,770.35	6,420,786	1,157,081	2351.03754	0.581632673	0.968572849
All Nippon airways (ANA)	6479	14,683,532	2,556,513.78	4,286,268	2,059,289	1113.020502	0.432154675	1.005479454
American airlines	23,102	34,707,729	8,654,892.94	17,866,791	2,417,898	2033.769025	0.584442993	0.944786721
British airways	16,563	21,401,581	5,304,411.47	10,079,586	4,438,214	2368.813,145	0.678351754	0.95344436
Delta air lines	17,408	27,292,425	7,349,946.47	14,571,329	1,671,083	2078.429,466	0.595125277	0.967574377
Emirates	13,153	27,369,447	4,717,271.61	11,276,662	6,531,110	4175.382,525	0.650644202	1.024964862
Garuda Indonesia	2187	2,834,184	676,346.53	1,514,745	282,129	1156.502,238	0.634000474	1.050834426
KLM	8101	15,090,771	3,027,818.18	7,347,192	4,093,466	1877.147,973	0.758122829	0.958613422
Lufthansa	33,288	27,007,957	5,759,785.56	12,398,774	6,928,900	1321.710,521	0.715628879	1.010927382

Malaysia airlines	5231	7,292,543	1,606,904.02	2,997,171	2,072,022	2183.9869	0.69512007	0.956114589
Qantas	12,156	17,368,244	3,156,052.26	9,945,797	2,623,457	1636.73531	0.723691698	1.015279346
SAS Scandinavian airlines	4046	4,152,670	1,108,178.33	2,304,528	344,994	815.6,428,041	0.638028545	0.934966136
Singapore airlines	9467	21,286,125	3,513,668.99	7,733,939	6,559,460	4646.417,801	0.67148901	0.960890949
TAM	6810	7,840,248	2,015,096.39	3,935,997	155,797	1201.092723	0.521895991	1.034039484
Thai airways	7374	10,441,041	2,417,856.19	4,725,671	2,157,255	2579.775,988	0.659218367	1.06012395
United airlines	18,460	29,065,589	7,647,835.29	14,645,900	2,340,509	2327.170,577	0.584416473	0.978280543

References

Assaf, A.G., Josiassen, A., 2011. The operational performance of UK airlines: 2002—2007. Journal of Economics Studies 38, 5—16.

Banker, R.D., Morey, R.C., 1986. Efficiency analysis for exogenously fixed inputs and outputs. Operations Research 34, 513—521.

Banker, R.D., Natarajan, R., Zhang, D., 2019. Two-stage estimation of the impact of contextual variables in stochastic frontier production function models using data envelopment analysis: second stage OLS versus bootstrap approaches. European Journal of Operational Research 278, 368—384.

Barbot, C., Costa, A., Sochirca, E., 2008. Airlines performance in the new market context: a comparative productivity and efficiency analysis. Journal of Air Transport Management 14, 270—274.

Barros, C.P., Peypoch, N., 2009. An evaluation of European airlines' operational performance. International Journal of Production Economics 122, 525—533.

Barros, C.P., Assaf, A., 2009. Bootstrapped efficiency measures of oil blocks in Angola. Energy Policy 37, 4098—4103.

Barros, C.P., Garcia-del-Barrio, P., 2011. Productivity drivers and market dynamics in the Spanish first division football league. Journal of Productivity Analysis 35, 5—13.

Barth, J.R., Caprio, G., Levine, R., 2013. Bank regulation and supervision in 180 countries from 1999 to 2011. Journal of Financial Economic Policy 5 (2), 111—219.

Bhadra, D., 2009. Race to the bottom or swimming upstream: performance analysis of US airlines. Journal of Air Transport Management 15, 227—235.

Bilodeau, D., Crémieux, P.-Y., Jaumard, B., Ouellette, P., Vovor, T., 2004. Measuring hospital performance in the presence of quasi-fixed inputs: an analysis of Québec Hospitals. Journal of Productivity Analysis 21, 183—199.

Bogetoft, Otto, 2011. Benchmarking with DEA, SFA, and R. Springer, New York.

Bogetoft, P., Otto, L., 2015. Benchmark and Frontier Analysis Using DEA and SFA. Package 'Benchmarking'. Version 0.26. https://cran.r-project.org/web/packages/Benchmarking/Benchmarking.pdf.

Chiou, Y.-C., Chen, Y.-H., 2006. Route-based performance evaluation of Taiwanese domestic airlines using data envelopment analysis. Transportation Research Part E: Logistics and Transportation Review 42, 116—127.

Choi, K., 2017. Multi-period efficiency and productivity changes in US domestic airlines. Journal of Air Transport Management 59, 18—25.

Coelli, T., Rao, D.S.P.,, Battese, G.E., 1999. An Introduction to Efficiency and Productivity Analysis. Kluwer Academic Publishers.

Coelli, T., Rao, D.S.P., O'Donnell, C.J., Battese, G.E., 2005. An Introduction to Efficiency and Productivity Analysis, second ed. Springer, Boston.

Cooper, W.W., Seiford, L.M., Tone, K., 2000. Data Envelopment Analysis. Kluwer Academic Publishers.

Daraio, C., Simar, L., Wilson, P.W., 2018. Central limit theorems for conditional efficiency measures and tests of the 'separability' condition in non-parametric, two-stage models of production. Econometrics Journal 21, 170—191.

Essid, H., Ouellette, P., Vigeant, S., 2010. Measuring efficiency of Tunisian schools in the presence of quasi-fixed inputs: a bootstrap data envelopment analysis approach. Economics of Education Review 29 (4), 589—596.

Fernandes, F.D.S., Stasinakis, C., Bardarova, V., 2018. Two-stage DEA—truncated regression: application in banking efficiency and financial development. Expert Systems with Applications 96, 284—301.

Golany, B., Roll, Y., 1993. Some extensions of techniques to handle nondiscretionary factors in data envelopment analysis. Journal of Productivity Analysis 4, 419—432.

Hoff, A., 2007. Second stage DEA: comparison of approaches for modelling the DEA score. European Journal of Operational Research 181, 425—435.

Hsiao, H.C., Chang, H., Cianci, A.M., Huang, L.H., 2010. First financial restructuring and operating efficiency: evidence from Taiwanese commercial banks. Journal of Banking & Finance 34 (7), 1461—1471.

Huang, C.C., Hsu, C.C., Collar, E., 2021. An evaluation of the operational performance and profitability of the U.S. airlines. International Journal of Global Business and Competitiveness 16, 73—85.

Imdadullah, M., Aslam, M., 2017. Multicollinearity Diagnostic Measures. Package 'mctest'. Version 1.1.1. https://cran.r-project.org/web/packages/mctest/.

Imdadullah, M., Aslam, M., Altaf, S., 2016. Mctest: an R package for detection of collinearity among regressors. The R Journal 8 (2), 499—509.

Jha, S., Hui, X., Sun, B., 2013. Commercial banking efficiency in Nepal: application of DEA and Tobit model. Information Technology Journal 12 (2), 306—314.

Kao, M., Lin, C., Hsu, P., Chen, Y., 2011. Impact of the financial crisis and risk management on performance of financial holding companies in Taiwan. World Academy of Science, Engineering and Technology 50, 413—417.

Kleiber, C., Zeileis, A., 2020. Applied Econometrics with R. Package 'AER'. Version 1.2—10. https://cran.r-project.org/web/packages/AER/index.html.

Lee, B.L., Worthington, A.C., 2014. Technical efficiency of mainstream airlines and low-cost carriers: new evidence using bootstrap data envelopment analysis truncated regression. Journal of Air Transport Management 38, 15—20.

Lee, T., Chih, S., 2013a. Does financial regulation enhance or impede the efficiency of China's listed commercial banks? A dynamic perspective. Emerging Markets Finance and Trade 49, 132—149.

Lee, T., Chih, S., 2013b. Does financial regulation affect the profit efficiency and risk of banks? Evidence from China's commercial banks. The North American Journal of Economics and Finance 26, 705—724.

McDonald, J., 2009. Using least squares and Tobit in second stage DEA efficiency analyses. European Journal of Operational Research 197, 792—798.

Merket, R., Hensher, D.A., 2011. The impact of strategic management and fleet planning on airline efficiency—a random effects Tobit model based on DEA efficiency scores. Transportation Research Part A 45, 686—695.

Mhlanga, O., 2019. Factors impacting airline efficiency in southern Africa: a data envelopment analysis. Geojournal 84, 759—770.

Mhlanga, O., Steyn, J., Spencer, J., 2018. The airline industry in South Africa: drivers of operational efficiency and impacts. Tourism Review 73, 389—400.

Müniz, M.A., 2002. Separating managerial inefficiency and external conditions in data envelopment analysis. European Journal of Operational Research 143 (3), 625—643.

Nemoto, J., Goto, M., 2003. Measurement of dynamic efficiency in production: an application of data envelopment analysis to Japanese electric utilities. Journal of Productivity Analysis 19 (2), 191—210.

Ouellette, P., Vierstraete, V., 2004. Technological change and efficiency in the presence of quasi-fixed inputs: a DEA application to the hospital sector. European Journal of Operational Research 154 (3), 755–763.

Pinheiro, J., Bates, D., R Core Team, 2022. Linear and Nonlinear Mixed Effects Models. Package 'nlme'. Version 3.1-160. https://cran.r-project.org/web/packages/nlme/nlme.pdf.

Pindyck, R., Rubinfeld, D., 1991. Econometric Models & Economic Forecasts. McGraw Hill, Inc., New York, NY.

Pitfield, D.E., Caves, R.E., Quddus, M.A., 2012. A three-stage least squares approach to the analysis of airline strategies for aircraft size and airline frequency on the north Atlantic: an airline case study. Transportation Planning and Technology 35, 191–200.

Ray, S., 1988. Data envelopment analysis, nondiscretionary inputs and efficiency: an alternative interpretation. Socio-Economic Planning Sciences 22, 167–176.

Ray, S.C., 1991. Resource-use efficiency in public schools. A study of Connecticut data. Management Science 37, 1620–1628.

Ruggiero, J., 1996. On the measurement of technical efficiency in the public sector. European Journal of Operational Research 90, 553–565.

Ruggiero, J., 1998. Non-discretionary inputs in data envelopment analysis. European Journal of Operational Research 111, 461–469.

Saranga, H., Nagpal, R., 2016. Drivers of operational efficiency and its impact on market performance in the Indian airline industry. Journal of Air Transport Management 53, 165–176.

Scheraga, C.A., 2004. Operational efficiency versus financial mobility in the global airline industry: a data envelopment and Tobit analysis. Transportation Research Part A 38, 383–404.

Simar, L., Wilson, P.W., 2007. Estimation and inference in two-stage, semi parametric models of production processes. Journal of Econometrics 136, 31–64.

Simar, L., Wilson, P.W., 2011. Two-stage DEA: caveat emptor. Journal of Productivity Analysis 36 (2), 205–218. https://doi.org/10.1007/s11123-011-0230-6.

Simar, L., Wilson, P.W., 2013. Estimation and inference in nonparametric frontier models: recent developments and perspectives. Foundations and TrendsVR in Econometrics 5 (3–4), 183–337.

Simar, L., Wilson, P.W., 2020. Hypothesis testing in nonparametric models of production using multiple sample splits. Journal of Productivity Analysis 53, 287–303. https://doi.org/10.1007/s11123-020-00574-w.

Simm, J., Besstremyannaya, G., 2020. Robust data envelopment analysis for R. Version 1.2-6. https://cran.r-project.org/web/packages/rDEA/rDEA.pdf.

Wilson, P., 2008. Frontier Efficiency Analysis with R. Package' FEAR'. Version 3.1. https://pww.people.clemson.edu/Software/FEAR/fear.html.

Wu, Y., He, C., Cao, X., 2013. The impact of environmental variables on the efficiency of Chinese and other non-Chinese airlines. Journal of Air Transport Management 29, 35–38.

Chapter 9

Conclusion

The aim of this book was to provide data envelopment analysis (DEA) practitioners DEA models written in the R programming script. By providing a guide on the use of DEA and the associated R scripts, and interpretation of the results, I hope it makes life much easier for DEA practitioners. I know this through my own experience as a DEA practitioner that it took significant time and effort to search for the appropriate DEA models and learn to apply them in efficiency and productivity analysis.

This book covers a number of DEA R scripts but is by no means complete. The purpose was to create a book that provides R scripts on key DEA models that would generally meet the needs of most DEA practitioners. Nonetheless, many of the DEA models in R script are available from the R packages, which this book relied upon. These include "Benchmarking" (Bogetoft and Otto, 2015), "deaR" (Coll-Serrano et al., 2018), "DJL" (Lim, 2021), "FEAR" (Wilson, 2008), "productivity" (Dakpo et al., 2018), "rDEA" (Simm and Besstremyannaya, 2020), and "TFDEA" (Shott and Lim, 2015). There are many more DEA models in these R packages that one could use to perform efficiency and productivity analysis. But for the purpose of this book, we relied on DEA models that best reflect and measure the performance of airline efficiency and productivity.

Each chapter (i.e., Chapters 3—8) showed the use of different DEA models and their purpose in measuring the efficiency and productivity of a sample of international airlines. For example, Chapter 3 covered standard DEA, and in Chapter 6, bad outputs (e.g., carbon emissions) were incorporated into the DEA model. This was followed by network DEA in Chapter 7. What this example highlights is that if we ignored carbon emissions as bad outputs (i.e., assume strong disposability), the standard DEA would be the appropriate model. However, carbon emissions is a by-product of airlines as discussed in Chapter 6, which suggests that carbon emissions cannot be ignored and that society faces a cost to clean up this bad output. In terms of NDEA, Chapter 7 showed that the airline industry comprises multiple phases—maintenance/servicing phase, operational phase and financial phase, which makes up the network structure of an airline. Although one could simply focus on the efficiency performance of the *delivery* portion of airline operations (i.e., using standard DEA), the efficiency measurement is incomplete and does not truly measure an airline's efficiency. This is because the key motivation of airlines is

Productivity and Efficiency Measurement of Airlines. https://doi.org/10.1016/B978-0-12-812696-7.00009-X

to remain efficient, which indicates that all segments of airline operations, both *provision* and *delivery*, must be considered in the production model. To complete the efficiency and productivity analysis, Chapter 8 examined various regression models to determine sources of efficiency.

From the sample analysis of international comparisons of airlines, we observe that for *delivery* and under VRS, many airlines were efficient. However under CRS, only KLM, Qantas, and Singapore Airlines were efficient, which suggests that these airlines' scale of operations are at the optimal level. The same results showing KLM, Qantas, and Singapore Airlines as efficient were also observed under various DEA models such as the bias-corrected DEA estimation, metafrontier, SBM, superefficiency model, and DDF. Same findings were also observed under cost-efficient, but under revenue efficiency, only KLM and Qantas were efficient. When bad outputs are considered, the number of efficient airlines decreases under VRS, and under CRS, KLM, Qantas, and Singapore Airlines remain efficient. Under NDEA, the findings differed from the previous findings. This is largely due to the impact node 1 (i.e., *provision*) has on node 2 (i.e., *delivery*) in an NDEA setting, which would impact on overall efficiency. The findings reveal that under CRS, Singapore Airlines remained the only efficient airline.

A final remark regarding measurement of efficiency and productivity is that the aforementioned analyses are based on a small sample size. Ideally, a large sample size would provide better and robust results in identifying efficient and inefficient airlines. But as mentioned in Chapter 3, the sample was determined by data availability, which could be used for all the DEA models for illustrative purpose. Finally, although not covered in this book, one could further merge the measurement ideas of Chapter 6 and 7 together to form NDEA incorporating bad outputs as noted in the literature in Chapter 2. Further extensions of this could also include a time series analysis, otherwise known as dynamic DEA.

References

Bogetoft, P., Otto, L., 2015. Benchmark and Frontier Analysis Using DEA and SFA. Package 'Benchmarking'. Version 0.26. https://cran.r-project.org/web/packages/Benchmarking/Benchmarking.pdf.

Coll-Serrano, V., Bolós, V., Benítez, R., 2018. Conventional and Fuzzy Data Envelopment Analysis. R package 'deaR'. Version 1.3. https://cran.r-project.org/web/packages/deaR/deaR.pdf.

Dakpo, K.H., Desjeux, Y., Latruffe, L., 2018. Indices of Productivity Using Data Envelopment Analysis (DEA). Package 'productivity'. Version 1.1.0. https://cran.r-project.org/web/packages/productivity/productivity.pdf.

Lim, D.-J., 2021. Distance Measure Based Judgment and Learning. Package 'DJL'. Version 3.7. https://cran.rstudio.com/web/packages/DJL/index.html.

Shott, T., Lim, D.J., 2015. TFDEA (Technology Forecasting using DEA). Package 'TFDEA'. Version 0.9.8. https://mran.microsoft.com/snapshot/2014-11-17/web/packages/TFDEA/index. html.

Simm, J., Besstremyannaya, G., 2020. Robust data envelopment analysis for R. Version 1.2-6. https://cran.r-project.org/web/packages/rDEA/rDEA.pdf.

Wilson, P., 2008. Frontier Efficiency Analysis with R. Package 'FEAR'. Version 3.1. https://pww. people.clemson.edu/Software/FEAR/fear.html.

Index

Printed in the United States
by Baker & Taylor Publisher Services